IN THE LIGHT OF
SCIENCE

IN THE LIGHT OF
SCIENCE

Our Ancient Quest for Knowledge
and the
Measure of Modern Physics

DEMETRIS NICOLAIDES

 Prometheus Books

59 John Glenn Drive
Amherst, New York 14228

Published 2014 by Prometheus Books

Cover design by Grace M. Conti-Zilsberger
Background cover image © Bettman/Corbis
Left and right cover images © Bigstock
Center cover image © Ian Cumming/Ikon Images/Corbis

Inquiries should be addressed to
Prometheus Books
59 John Glenn Drive
Amherst, New York 14228
VOICE: 716–691–0133
FAX: 716–691–0137
WWW.PROMETHEUSBOOKS.COM

18 17 16 15 14   5 4 3 2 1

Library of Congress Cataloging-in-Publication Data Pending

Printed in the United States of America

Dedicated to
my wonderful daughter, Maria-Christina, who is curious about the world,
my loving wife, Anna,
and the memory of my beloved father.

CONTENTS

8   CONTENTS

ACKNOWLEDGMENTS

I am always grateful to Professor Alexander A. Lisyansky, my PhD thesis mentor, who taught me physics and the art of fundamental research. Many thanks to Professors Dennis Organ and Ivana Djuric for their willingness to review the manuscript at its early stages and for their helpful comments and suggestions.

I would like to thank my agent, Nancy Rosenfeld, who believed in my project and has stood by me with useful advice. I have benefited immensely from her experienced recommendations.

I am grateful to Bloomfield College for giving me the opportunity to research this topic and also to my students whose thirst for knowledge keeps me improving.

I thank my parents for their unconditional love and guidance throughout the years. I am grateful to my wife, Anna, for encouraging discussions over coffee early, early in the morning during all the stages of this book project. Her insightful comments led to several improvements.

I am thankful to the staff of Prometheus Books for all their help, including Jade Zora Scibilia, Julia DeGraf, Mark Hall, and Melissa Raé Shofner. I am deeply indebted to my gifted editor, Steven L. Mitchell, for his thorough recommendations, perceptive questioning, and generous editorial assistance, which, without a doubt, have contributed considerably to the enhancement of the manuscript.

But most of all, I am thankful to a nine-year-old, my daughter, my pride and joy, my Maria-Christina, whose curious mind and love for knowledge kept me focused. Anna and our daughter, Maria-Christina, mean the world to me. Thank you!

PROLOGUE

THEME

In an attempt to discover the roots of science and understand the development of our scientific knowledge about nature as a series of logical progressions, this book narrates a concise history of humans from the "beginning" (from when *Homo sapiens* evolved two hundred thousand years ago), by investigating, in the light of science, subtle interconnections among the three most significant cultural landmarks of humanity: (1) the culturally explosive urbanization, ten thousand years ago, and the unavoidable birth of civilization gradually thereafter—phenomenal events that triggered a wealth of new pursuits including religion; (2) the intellectually revolutionary birth of science, 2,600 years ago, which broke the bonds of superstition; and (3) the scientifically extraordinary modern era of quantum physics, relativity, string theory, and "the God Particle"[1]—of mind-boggling science that contributed to a better understanding of the universe and skyrocketed progress in technology, although it also challenged society.

I trace the intellectual continuity in our efforts to know nature by describing human life when it was primitive, outlining the pivotal transition in lifestyle from nomadic hunting-gathering to settled urban communities (i.e., to the birth of civilization), discussing the life-changing birth of religion and its influence on the eye-opening birth of science, as well as the intellectual evolution from mythology to science and its causes, thus offering us the first scientific theories, conceived during the sixth and fifth centuries BCE, analyzed from the context of modern science.

It is a book about the history of science and about science itself both ancient (particularly pre-Socratic theories) and modern (particularly physics) and in light of each other. This comparative approach to science demonstrates that the measure of modern physics is both an expansion and a reflection of various scientific ideas conceived by prominent pre-Socratic philosophers. And also that scientists of the twenty-first century are still grappling with the fundamental problems they raised some twenty-five hundred years ago.

In the Light of Science begins with its more historic part I, and proceeds to its more scientific part II, by blending history, religion, philosophy, science, and overall human culture. The theme of this book is discussed more analytically in its two parts.

PART I: FROM CHAOS TO ORDER

An epoch of sheer struggle for survival began with the evolution of *Homo sapiens* about two hundred thousand years ago. This was followed by the first form of clear self-expression through cave painting around thirty thousand years ago. About twenty thousand years later came the domestication of animals and plants, which led to herding, agriculture, the cultural phenomenon of urbanization, and the consequent birth of civilization. With civilization came the remarkable birth of religion and the development of mythological worldviews, events all of which were starting to occur roughly ten thousand years ago. Soon there would be complex, busy city-states, some four thousand years later, followed by the momentous invention of writing and the beginning of written history, a few centuries later. The enlightening birth of science began about 2,600 years ago.

Overall, this part presents a brief history of humans with the goal of understanding the main events leading to the most critical transition in the evolution of human thought, the shift from mythology to science—from the mythological and apparent worldview of deceptive senses to the rational yet intangible worldview of inventive intellect.

Contributing to this transition was the rise of the Greek civilization. About 2,600 years ago the ancient Greeks had a magnificent intellectual awakening. "Suddenly" the age-old popular mythological worldviews were questioned, rethought, and eventually changed. Nature was no longer seen as a chaos of random, unpredictable, and incomprehensible phenomena attributed to mysterious supernatural forces through myths, superstition, and the chancy decisions of capricious, anthropomorphic gods. On the contrary, nature was viewed as a cosmos: a well-structured, organized, ordered, harmonious, self-contained, self-consistent, and beautiful whole in which the phenomena were *natural* components that obeyed intrinsic causal laws that could be discovered and understood by the practice of rational analysis of nature and without invoking the supernatural. A profound transition in human thought took place that was a consequence of the realization by these Greeks that nature *is* comprehensible. A simple question emerged: what is the nature of nature? The prolific answers Greek thinkers offered ascribed purely naturalistic causes to all phenomena in nature and gave birth to science.

This part examines the birth of science and discusses why Greece may be considered its birthplace. I argue that while the generally accepted conditions (those dealing with geography, economics, religion, and political structure) might have been necessary for the rise of Greek civilization and particularly for the birth of science, they were hardly sufficient. Thus a hypothesis will be proposed: that these together with three other conditions—(1) the influence of the evolving ancient Greek language, (2) the force of intellectual habits, and (3) the intriguing ancient Greek idiosyncrasy—all had a critical role to play.

The factor of language has not been sufficiently appreciated, but neither has the unusual factor of intellectual habits, through which I hope to explore how the phenomenon of natural selection from biological evolution has contributed to the birth and constant development of the scientific outlook at the expense of the mythological one. Moreover, the factor of Greek idiosyncrasy is often overlooked, but I believe that any effort to understand the rise of a civilization is probably incomplete without also an attempt to understand the idiosyncrasy of the people who caused such a rise.

The earliest scientific theories were formulated by the pre-Socratics, the Greek natural philosophers from the sixth and fifth centuries BCE who were the first (at least that we know of) to explain the phenomena of nature solely in terms of naturalistic causes and consequently to contribute significantly in the early genesis of science. That said, there is absolutely no reason to believe that, before or after the Greeks and independently of them, others in the world would not have conceived the scientific interpretation of nature. In fact, the proof of my claim is that all kinds of people from all over the world do, or can learn to do, science. Nonetheless if something profound like practicing science—that is, seeking exclusively *naturalistic* (scientific and rationalistic rather than supernatural) explanations for *all* the phenomena of nature, including the origin and evolution of humankind and of life in general—had not happened completely, systematically, freely, enthusiastically, and most importantly, truly *consciously* (not accidentally or merely instinctively) and *habitually* (not occasionally), as in the case of the pre-Socratics, such good skill would not have evolved into a cultural tradition, would not have spread, and almost certainly would have quickly vanished without demonstrating a significant effect on society. But evolution occurs when a phenomenon leaves a mark on the environment. The first scientific mark, which served as a basis for science thereafter, was left by the ancient Greeks. In this perspective science may be regarded as having been born out of the Greek civilization. Nevertheless, from a grander point of view the story of science—seeking the nature of nature—begins with the evolution of the human species two hundred thousand years ago. For, out of the myriad species that have existed, only humans managed to develop a rational interpretation of nature. Why that was so is seen here too where the main events leading up to the birth of science are discussed through a brief history of humans.

PART II: THE PRE-SOCRATICS IN LIGHT OF MODERN PHYSICS

In an effort to trace the beginnings of science, this part introduces the most important scientific theories of the most famous pre-Socratics (including Pythagoras and Democritus) and analyzes them within the context of modern physics (or science in general). These philosophers and thinkers knew that nature is intelligible. It has secrets, to be sure, but like Promethean fire, their knowledge is discoverable and can be harnessed. They had ideas that were so phenomenal, fascinating, unique, strange, and daring that some actually anticipated various aspects of modern science that were well ahead of their time. And some of these ancient ideas, while defying common sense and apparent reality, have not yet been refuted—they remain still unsolved mysteries! These Greek theories are not as obsolete as is often assumed. They still spark the imagination. Hence our intellectual trip exposes the most beautiful and mind-blowing laws of nature and shows that, despite the two-and-a-half-millennia time difference, ancient and modern science share a fundamental qualitative similarity more often than is usually thought, and they complement each other's scientific uniqueness.

Modern physics is becoming more and more complex, both conceptually and mathematically, often hitting intellectual dead ends. But while the laws of physics are mathematical, its mathematics emerges from its concepts. And the pre-Socratics were undoubtedly masters of profound conceptual analysis (of course, some were also excellent mathematicians who conceptualized their physical theories using the mathematics known at the time). Therefore, ancient rationales will at times be used to reexamine and reassess some of the fundamental premises of current theories of physics. And while one of our basic questions is how ancient science measures up to modern physics, we may often find that the question can be reversed: how does modern physics measure up to ancient science?

If the Greek philosophers (i.e., Socrates, Plato, and Aristotle) are correct and philosophy truly does begin in wonder, and if natural philosophy is the precursor to what we now know as science, then it makes perfect sense that the urge to ask questions and to seek answers is a critical part of the scientific state of mind and its rise from the darkness of superstition to the light of knowledge.

Throughout its history science has asked many questions: What is the nature of nature? Does nature obey laws that can be understood? If yes, how do these laws work?

What are the things (stuff, matter) all around us made of? Is there a single primary substance of matter? What is the nature of matter? Does matter's behavior depend on space and time, and vice versa? What is the nature of space and time?

Is the universe finite (in size and age) or infinite? Is it static and fixed, expanding, contracting, or vacillating? Does it go through cycles? Is there some one thing that caused everything that now happens—a first cause? Was there a specific beginning of time? Will there be a definite end?

Where did the species, including humankind, come from? Is there a biological development process?

Is reality objective or subjective? Is this the only universe, or are there also others, each with its own unique reality? Can the intellect alone conceive a truer reality, or does it need the aid of the senses?

Is absolute knowledge ever attainable? Does God exist? Can science prove or disprove the existence of God, or will God always remain a matter of subjective belief? Why are humans intelligent? In fact, why are we the most intelligent of all the living or extinct species we know?

What is the nature of nature? It's a deceptively simple question but one that has packed within it the many additional questions outlined above—and many more besides. Two and a half millennia ago this fundamental question captivated the minds of the ancient Greeks and led them to strange but *rational* answers that demystified and demythologized nature and gave birth to science. From that time on, science has been influencing the world by guiding it out of the cave of ignorance and into the light of truth.

PART I

FROM CHAOS TO ORDER

CHAPTER 1

PLATO'S PARABLE OF THE CAVE

INTRODUCTION

The cave parable in Plato's *Republic* elegantly captures the essence of nature and the goal of scientists: nature has its secrets, and it is the role of scientists to discover, or "steal," them, as Prometheus stole divine fire from the gods and gave it to humanity.

Apparent reality is as incomplete and nebulous as the shadows of objects in the parable. The prisoners who live in the cave and see only shadows on the walls consider only these shadows to be real. They have no knowledge, none at all, of the objects that cast the shadows. But the prisoner who escapes from the cave has at last a better sense of reality and discovers for herself that the shadows are only a mere copy of the real objects. She sees that things are more than they appear. The parable helps us to realize that nature is much more beautiful and complex than the way the senses alone display it for us. And so for the scientist (as for the prisoner who escapes the illusive shadows of the cave) nature *is* comprehensible through both observation and rational contemplation. Logic unveils a much richer reality than the tangible one of mere sense perception alone. Ever since science opened the door to a rational understanding of nature, its discoveries have continued to amaze us by defying both common sense and apparent reality.

THE PARABLE

Imagine an underground cave where prisoners are held in chains since childhood and for many years thereafter, restricted from moving or even turning their heads, but are only able to look straight ahead and toward a wall in front of them. Behind and far from them is a fire the light of which projects on the wall shadows of various objects (including those of themselves, of passersby behind them, and of other things in general). Because of how they are bound, the prisoners see nothing else but these shadows. They believe that only these shadows are real, and the prisoners are utterly ignorant of the objects that project them, or generally of the greater truth about nature that is indeed accessible if only they could escape.

One day a prisoner does escape. Her instinctive curiosity guides her away from the darkness of the cave, first toward the fire and then upward and out of the cave and into the light of the sun. But being so long in the dark she is not accustomed to the light. And so she is frightened by her inability to see clearly. Gradually her eyes adjust, and she sees things she has never seen before, a panorama that is grand, boundless, magical. She is startled but also puzzled. Nature is so much different from the way she had imagined it. For the first time she sees mountains, flowers, people, animals; she sees the blue sky and the sea and the colors in the trees. She observes but also contemplates. She notices how similar all objects are to their shadows. And then for a moment she seems lost, anxious, absorbing all that is around her. While deep in thought she suddenly exclaims eureka! It is an exhilarating moment. She has discovered something. At last she has a better sense of reality because she realizes that the shadows in the cave are just mere copies of the real objects. Nature is more than it appears. But it *can* be understood!

Sad for her friends and morally obligated to free them from the deceptive darkness, she descends once more into the cave. But upon reentering she is frightened for a second time, for again she cannot see: her eyes have now adjusted to the light and are no longer accustomed to the dark. This does not go unnoticed by her friends, who think she is now different from them, that she has changed, that her vision is destroyed by the light. They think *she* is blinded by the *light*, not realizing it is *they* who are blinded by the *dark*. Her every effort to convince them otherwise and free them is futile and mocked. In fact any new knowledge she may impart is viewed as a threat, and they contend that anyone claiming to have such knowledge should be put to death.

MORALS

To know nature as it *really* is, we must use both our senses and logical reasoning. Sense perception alone is like the cave reality, incomplete but not absolutely false. What is false are the generalizations we make by relying solely on the experiences of the senses. Shadows *are* real, but not the only real things; shadows are *not* the truth, just *part* of it. Freeing oneself from the dark bondage of the cave in order to enter the light symbolizes the transition from a world of apparent reality, ignorance, and superstition to one of rational thought and potential enlightenment. In achieving such an escape, fear of learning (the pain in the eyes caused by the light) is eventually overcome, and superstition is replaced by reason.

The parable is also a story about ignorance and the cave-dwellers' failure to recognize their ignorance. Not only do the prisoners lack knowledge, but they do not know that they lack it. They were prisoners of time-honored prejudices and illusive sense perceptions. The cave may be seen as a metaphor for the incomplete knowledge we have about everything and the dangers associated with that partial understanding; especially when we do not realize that our knowledge is only partial, and therefore we think we know more than we do. Leaving the cave (to be educated) is as difficult as entering it to make others aware (to educate). The latter is far more advantageous because one is at least equipped with some knowledge that can be utilized to possibly anticipate forthcoming challenges.

CONCLUSION

Reading a symbolic version of a very complex story as if the symbols were real and the symbolic version were the actual story is an inherent risk in the very symbols used. Sense perception by itself is telling us just a symbolic story of nature—we see only shadows or imperfect representations of reality. But when combined with our intellect the two offer hope of providing us with a more realistic version of the story of nature. The pre-Socratics realized early on that appearances deceive, that the sensible world is like a cave, both illusive and incomplete. But even though our senses can be deceived, our rational intellect is perceptive and so offers a much-needed check on the veracity of what our senses report. This has helped us advance from a strictly sensual worldview to one in which our reason carefully assesses and analyzes what our senses convey about the world beyond us. And these ancients found that nature is a lot more intriguing, mysterious, and beautiful than it may at first appear, but also that it is intelligible to the willing mind only through the careful and deliberative approach of science.

CHAPTER 2

WHAT IS SCIENCE?

INTRODUCTION

Science (taken from the word for knowledge in Latin) is the systematic study of nature and the organization of acquired knowledge into timeless, universal, causal, and, most importantly, testable laws and theories that are derived from observation and rational consideration. A good scientific theory, therefore, makes experimentally verifiable predictions whose confirmation leads to the theory's acceptance as a true description of nature. The premise of science is the realization that with rational thought nature *is* comprehensible. Science holds that natural phenomena, while often appearing to be random and unpredictable, are actually orderly and to a certain degree predictable; they obey intrinsic causal laws that can be understood rationally without the need of invoking myths, superstition, supernatural forces, or the intervention of capricious and anthropomorphic gods.

CAUSALITY

Scientists are generally viewed as rational people, but like the rest of us they are passionate and willing to sacrifice much in the pursuit of their endeavors. Democritus once said that he would "rather discover one causal explanation than obtain the whole Persian empire."[1]

Causal explanations conform to the principle of causality, according to which later events are caused by earlier events. Causality is a relation between *causes* and *effects* that organizes knowledge in the hope of discovering general laws of nature. But the relation between cause and effect is still very much an open question. In a deterministic interpretation of nature a specific cause produces a specific effect. Say cause-1 produces effect-1. But cause-1 may also have a cause of its own—some previous cause, say, cause-2, which may itself be caused by cause-3, and so on. In fact the original effect-1 may also be the cause of some other effect, say, effect-2, which in turn may be the cause of effect-3, and so on. We can show this schematically as

$$\cdots \to \text{cause-3} \to \text{cause-2} \to \text{cause-1} \to \text{effect-1} \to \text{effect-2} \to \text{effect-3} \to \cdots$$

On the other hand, in a probabilistic interpretation of nature one of several probable causes could produce one of several probable effects. Therefore, it is obvious that neither causes nor effects are known with certainty; on the contrary, all events are expressible in terms of probabilities.

Classical physics (as understood by Isaac Newton's and Albert Einstein's theories) is deterministic, whereas quantum physics is probabilistic in its approach. The location, speed, direction of large objects (such as cars, billiard balls, and planets) can be described with accuracy by the laws of classical physics, but the same characteristics of tiny particles (such as electrons, protons, neutrons, neutrinos, quarks, etc.) are describable only in terms of probabilities through quantum physics. We can pinpoint the position of a car as well as its speed and direction, but this is not the case with an electron. Quantum theory can describe only where an electron is *likely* to be, not where it is or will be. The source of uncertainty and probability will be introduced in chapter 12, "Heraclitus and Change."

Causality only explains later effects by earlier causes, but it cannot explain the very first (or primary) cause; it can only assume the truth of such a cause and proceed from there to deduce its actual (or possible) effects. No cause can be assigned to the label "primary cause" since if it could, it would not be primary. Nevertheless, to find any causes and their effects, even the elusive initial cause, the phenomena of nature must be studied rationally with a method, the scientific method.

THE SCIENTIFIC METHOD

Isaac Newton (1642–1727) hypothesized that white light is a blend of all the colors of the rainbow. He tested his hypothesis by experimenting: he passed a beam of white light through a glass prism and observed how it disperses into a rainbow of color. Others, at Newton's time, thought that the colors came from the prism itself, that they were not a property of light. Newton proved them wrong by advancing his experiment. This time he passed just one color, say, the blue (from those dispersed by the first prism), through a second prism. Since only blue entered the second prism and only blue emerged from it (and not a rainbow of color), color, he concluded, was a property of light and not of the prism. He supported his hypothesis even further by mixing again the initially separated colors and noticing how once more they reproduced white light. The so-called Snell's law calculates the exact angle that each color of light refracts by crossing from the one medium, air, into the other, the glass prism. Hence what we must also emphasize here is that all scientific laws are quantified through mathematics. Without such quantification modern technology (of computers, cell phones, etc.) would not have been possible. Chapter 11, "Pythagoras and Numbers," will discuss how one may model nature mathematically.

As seen through the above example, the scientific method consists of observation, reason, and experimentation. By observing a natural phenomenon and thinking upon how it may have come about, its explanation is attempted by means of a rational hypothesis, a proposed explanation offered as a possibility. The validity of the hypothesis must be tested experimentally. A set of conditions is predicted such that if the hypothesis is confirmed, that set of conditions would in fact occur. A conclusion is reached about whether or not a hypothesis is verified (confirmed) or falsified (disconfirmed) by analyzing the predictions of the hypothesis in light of what is found to be the case when the hypothesis is tested. If the hypothesis is consistently verified by numerous individuals who attempt to test it, then a law (or a general theory of nature) is discovered and our comprehension of nature expands a bit more. Of course, often a hypothesis on how some aspect of nature works can be formed by pure theorization (reason) without first having to observe some particular phenomenon (as is the case of Einstein's theory of special relativity), although it is still likely that various prior observations would guide a scientist's speculations, for, in truth, no person is disconnected entirely from the influences of nature. The fewer assumptions a law makes and the more phenomena it explains, the more powerful it becomes. Furthermore, regardless how abstract and complex a law may be, it must be able to explain the natural phenomena as we see them, as we perceive them through our five senses. It is always possible, as any good scientist knows, that any law might later be shown with additional testing to be just a mere copy, a shadow, of some greater truth yet to be uncovered. In fact, this has been so throughout the history of science. Nevertheless while our scientific models may in the end prove to be nothing more than shadows of yet undiscovered truths, they still play a vital role in the development of human thought and understanding about the world and our place in it.

CONCLUSION

Science has not always been part of human civilization. In fact, even civilization has not always been part of human existence. So what critical event in human history created the potential for us humans to develop abstract thought and eventually science?

CHAPTER 3

URBANIZATION

INTRODUCTION

With perhaps the exception of the mastery of fire, the most momentous event in the roughly two-hundred-thousand-year history of *Homo sapiens* was urbanization, which emerged about ten thousand years ago. Had it not occurred, we would still be hunting and gathering. But it did, and that led to the birth of a new and complex way of life called civilization with everything this term entails. This development was the result of two great ideas that were beginning to be implemented at roughly the same time by our hunter-gatherer ancestors: namely, the domestication of animals, which led to herding, and the domestication of plants, which led to agriculture. Both activities, but particularly agriculture, promoted the need of a settled life, created food surpluses, and with these surpluses came an expanding population. From small villages of mere hundreds to crowded city-states of many thousands, life became increasingly more communal and diversified, an unprecedented transformation that by about six thousand years ago found the cultural phenomenon of urbanization spreading worldwide. To put these ideas in perspective, this chapter will narrate a brief history of *Homo sapiens* (covering our primitive beginnings and the main changes that occurred thereafter) from the time we first evolved two hundred thousand years ago to May 28, 585 BCE, the day of the solar eclipse predicted by Thales,[1] the first natural philosopher. Because Thales, a pre-Socratic, flourished around the time of this eclipse, this day may be regarded as the birthday of science. In fact, the entire sixth century BCE was an intellectually explosive period globally. But let's start from the "beginning."

THE ERA OF PURE SURVIVAL: 200,000–30,000 YEARS AGO[2]

Homo Genus[3]

The *Homo* genus consists of several member species all closely related to us, each of which is also unique. Its first member, *Homo habilis* (a distinct species having some humanlike characteristics), evolved more than two million years ago and became extinct about 1.4 million years ago. Its only extant member, *Homo sapiens*, evolved about two hundred thousand years ago in Africa and is the species to which every living human belongs—so each one of us can be traced back to a common African ancestor. *Homo sapiens'* evolution was happening just as our older evolutionary relative *Homo heidelbergensis*, who had evolved about seven hundred thousand years ago, was becoming extinct. It is at this point that our species might well have gotten its chance to evolve as a consequence of the eventual extinction of another species, in general not an uncommon phenomenon in the evolution of life. Sadly, the longest-living member of the genus *Homo*, *Homo erectus*, whose members evolved approximately 1.9 million years ago, also became extinct about 143,000 years ago.

Since their earliest days and down to about the end of the Paleolithic Age (the "Old Stone Age") roughly ten thousand years ago, *Homo sapiens* were basically hunter-gatherer nomads with remarkably few cultural changes. They constantly searched for food by following migrating wild animals and by seeking new areas of fresh edible plants. They lived and traveled in small social groups, including perhaps tens or hundreds of individuals from an extended family. Their diet included plants, nuts, eggs, honey, fruit, and meat from animals, fishes, and birds. Survival was a day-to-day affair, but they got by despite the dangers and uncertainties of their rough and wandering lifestyle. It is not easy to hunt large wild animals (horses, reindeer, mammoths, cattle, woolly rhinoceros, bears, and the like) with primitive weapons/tools. And food had to be found daily, but there was no guarantee of that. In fact *Homo sapiens* came near to extinction sometime between ninety thousand to seventy thousand years ago. During this period the climate had fluctuated significantly, and it was around this time, 74,000 years ago, that super volcano Toba erupted, complicating further the conditions for survival.[4] In spite of these challenging circumstances, *Homo sapiens* have been managing to survive.

Bottleneck Effect

It is speculated, however, that the human population may have suddenly decreased dramatically (to only a few thousand) at this time,[5] and as a consequence so may have the gene variety among the survived members—one of the reasons humans today are genetically so similar, hence our differences are truly skin-deep. To explain this genetic similarity biologists often employ the so-called bottleneck analogy. Imagine a narrow-neck bottle filled with marbles of many different colors. The marbles represent the initial large genetic variety in the population of a species. And the narrow neck represents a potential challenge that a species might have to experience—such as severely cold or hot weather, earthquakes, volcanic eruptions, droughts, earth-asteroid collisions. Now, just as the color variety of the few marbles that make it through the narrow neck (that "survive") is likely to be reduced when the bottle is turned over, the gene variety of the few members of a species that survive a harsh situation or catastrophe is also likely to be reduced.

The Peculiar Water

While extreme coldness is one of the toughest bottleneck passages a species might have to endure, fascinatingly, a life-preserving peculiar behavior of water eases the passage (lessens the difficulty). What is it?

Oceans, speculated to have been the habitat where primitive microscopic life-forms first evolved, begin to freeze from the surface downward, but they don't freeze all the way to the bottom when the temperature drops to freezing. The water at the bottom of an ice-covered ocean never freezes; it remains always around 4 degrees Celsius (about 39 degrees Fahrenheit). This is because of an unusual behavior of water: when cooled below 4 degrees it expands rather than contracts like most other materials. That's why a water bottle forgotten in a freezer will break—as the water in it cools below 4 degrees it expands (especially so as it turns into ice at 0 degrees Celsius) and cracks the bottle. Similarly, ice floats because its expansion at lower temperatures also lowers its density below that of liquid water, causing it to rise to the surface.

Now the water at the bottom of an ice-covered ocean can't lower its temperature below 4 degrees because it literally does not have the energy to do so. For it to cool below 4 degrees water must expand, that is, it must lift up all that heavy icy sheet over it, but it can't. Similarly, we can't lift something extremely heavy over our heads—we don't have the strength or the energy to do it. The water at the bottom of an ice-covered ocean remains nice and cool at 4 degrees, fluid, unfrozen, keeping the fish swimming, preserved, and pleased. Unlike deep oceans or lakes, however, water in shallow street pools does freeze completely because the quantity of the water in it is small; the water at the bottom of a pool can freeze as it easily lifts up the small weight of the water over it.

If water didn't possess this peculiar property, the evolution of life on earth might not have been possible at all because oceans would freeze *completely*, killing all life-forms within them. If water were contracting, not expanding, below 4 degrees, it wouldn't have to lift up any water or ice positioned above it in order to cool down during freezing weather. Instead it would easily become colder and transform into frozen solid ice first at the ocean's surface. Then, being denser than the water below (because of the assumed contraction) this ice would naturally sink to the bottom of the ocean (like a rock does because it's denser than water), and the water below would in turn be displaced upward toward the ocean surface. The new surface water would follow the same sequence of events (it would freeze, condense, sink, and displace the water below it to the surface) until all ocean water would freeze completely crushing to death all marine life. Even worse, in persistently cold weather life might not have gotten the chance to even evolve in the first place; the crushing forces of solid ice spreading everywhere in a frozen ocean would have been destroying the molecules of life, killing any chance they might have had to coalesce and develop. Thankfully, water is peculiar, and so life was able to get its start on earth.

Primitive Technology

The dawn of technology, marking the beginning of the Paleolithic Age, occurred more than two million years ago with the use of simple stone tools by *Homo habilis*, "handy man," or possibly by other species before him from the genus *Australopithecus*. *Homo sapiens*' tools, at different periods during the years covered in this section, included stone projectiles, blades, flakes, stone or bone hammers, long pointed wooden spears, stone or bone awls, needles and scrapers to perforate and turn hides into clothing, and in general anything made of stone, wood, or bone that could be useful for a hunting-gathering lifestyle amid varying and challenging climates. Metal tools and pottery did not yet exist.

Technology is often associated with science, but technology itself is not science. Tools and implements—primitive forms of technology—can exist without science. But usually, especially nowadays, technology is a consequence of science; it is the application of the laws of science for various practical purposes. Technology can also lead to science (as we will see in chapter 6, "The Birth of Science"), but it is not itself science.

As nomads who lived in small social groups, humans shared food and other resources, lived mostly outdoors but sometimes avoided the wind and cold weather of the last glacial period (110,000–10,000 years ago) by seeking refuge in natural shelters such as caves or human-made simple huts constructed by piling up stones or branches or by hanging animal hides on poles, or by gathering around a campfire.

Fire

The use of fire was a skill first practiced by *Homo erectus* (another of our extinct ancestors known for standing upright rather than crouched or bent over) possibly five hundred thousand years ago. Whether *Homo sapiens* discovered such skill independently or learned it by observing and imitating *Homo erectus'* example is uncertain. Also uncertain is when fire actually came into being as a human tool (see next section). An interest in fire may have been sparked in a way by something like this: after a naturally caused fire (e.g., as a result of lightning), a keen and courageous observer might have retrieved a burning branch, kept it ablaze by feeding it with new branches, and experimented with its properties.

Fire was used for warmth, for light at night, to keep insects and dangerous animals away, for hunting animals, and for cooking food. Cooking killed parasites (especially those on meat), made food softer and tastier, and increased the types of food that could be consumed. Cooking has aided in safer food consumption and digestion and has improved the survival chances of our species, and in so doing it has contributed to our general evolution and in particular to the evolution of the human brain. Eating cooked food delivers more energy than eating the same food raw.[6] This fact has been especially significant for the evolution of our energy-hungry complex brain, which although comprising only 2 percent of our body weight requires 20 percent of the energy from the food we consume.

Gathered around a campfire to relax, socialize, and bond, *Homo sapiens* might have even "told" their first stories, an activity that has certainly influenced the development of human language. I imagine these stories were very simple ones, language being in its most primitive form at that time. Regrettably, sounds do not fossilize, so before the first written record, left sometime during the fourth millennium BCE, the level and extent of language used by humans is uncertain. It is hypothesized that a form of basic vocal language must have evolved by one hundred thousand years ago.

The warmth of fire might have also helped parents to care for their helpless babies. Remarkably, among the primates, human babies require the longest period of parental care. By comparison, the childhood of chimpanzees, which genetically are our closest relatives from the living species, is roughly half as long as the childhood of humans. Caring for each other, conveying stories, and exchanging information undoubtedly enhanced *Homo sapiens'* survival skills and their evolutionary path. Moreover, learning from each other along with the long period of parental care contributed to the overall bond between humans. Parents who care for their own offspring, especially when this is done for a period of several years, instill in their children the tendency of caring as they are growing. As this process continues throughout the group and through generations, chances are that offspring and parents alike will also begin caring for their close family as well as their extended family. Such feeling could in turn inspire the desire for a more communal lifestyle, which is the basis for civilization. Directly or indirectly the phenomenon of fire has contributed to all aspects of the evolving human culture.

Migration

Between eighty thousand to sixty thousand years ago, *Homo sapiens* migrated from the African savannah first to Asia, then to Australia (ca. fifty thousand years ago), to Europe (ca. forty thousand years ago), to the Americas and to the Pacific islands (ca. fifteen thousand years ago), and by about ten thousand years ago even to the Arctic, but not to Antarctica. Migration was possible because the icy-cold temperatures of the last glacial period bound the water as ice on land, and so it could not flow into the sea. As a result, the sea level then was lower than that of today by about 100 meters, exposing more parts of the continental crust, which in turn functioned as natural land bridges between continents. These bridges lasted until about ten thousand years ago. Then the glacial period ended. The melted water flowed into the oceans and caused the sea levels to rise and these bridges to be submerged under the water. Consequently, the new islands that were created kept their inhabitants isolated.

So, by thirty thousand years ago *Homo sapiens* had developed better tools, built primitive huts, advanced their hunting and carving techniques, made and wore clothing, socialized around a hearth, buried their dead, possibly even played music (using flutes made from bones). But perhaps the most significant of their activities, which was not related to sheer survival, was painting! It was the first form of pure self-expression.

FROM PAINTING TO HERDING: 30,000–10,000 YEARS AGO

Painting

Art begins in prehistory more than thirty thousand years ago by our Paleolithic ancestors, the Cro-Magnons, who decorated their cave dwellings with splendid, colorful, and expressive animal paintings. Cro-Magnons, named so after the site in France where their remains were first discovered, are the *Homo sapiens* who arrived in Europe about forty thousand years ago and are believed to have looked and behaved more like modern-day humans. With their beautiful art they demonstrated their sensitivity and sensibility. Cave painting is "great art, manifested in works of subtlety and power that have never been surpassed."[7] While the true significance of these paintings may never be known, it has been speculated that in addition to painting for merely the sake of art, the animals of the paintings might have also had a sacred function.[8] As if life would imitate art and, as depicted in the cave walls, humans could become better hunters and wild animals captured and tamed. But if humans could express themselves so well through painting, they could probably speak as well, too. And so by now their stories around their hearths would have been more descriptive. Consequently they understood and learned from each other, making stronger both the human bond and their social lifestyle. Urbanization and the birth of civilization were thereafter naturally expected. Painting is a significant activity in human culture, for after painting comes writing—recorded information! This sequence of events is normal. For in painting one draws to represent things that already have a shape, for example, a bison. On the other hand, in writing, especially with an alphabet, one draws (e.g., letters) to represent things that do not have shape, that is, the sounds of a language. Comparatively, then, an alphabet is a more abstract invention than painting or a pictographic type of writing, and so it naturally followed both of them. How the Greek alphabet aided the pre-Socratics in conceiving the scientific view of nature is a topic to be contemplated in the chapter "The Birth of Science." A culture, even an urbanized one, cannot advance without recorded information. But between painting and writing there was a long time period and a few other significant cultural events, the first of which might have been lighting fire at will.

Fire Lit at Will

One of the most consequential events since our species evolved in Africa some two hundred thousand years ago was the lighting of fire at will. How and when this happened is unknown. But heat from friction is certainly what caused it. In their attempt, for example, to sharpen their tools by rubbing or striking them together (e.g., stone with stone or wood with wood), our curious ancestors could have easily started a fire. By rotating back and forth a pointed stick in a concavity of another in an effort to sharpen it further, a spark might have been ignited and fire lighted. An event this remarkable would not have gone unnoticed. It would have prompted the realization that fire can be lit and controlled at will! And although quite possibly that was an accidental discovery, lighting fire from that moment on became a favorite intentional human activity without which the notion of civilization would be nonexistent.

Fire has ever since been playing a critical role in our cultural and biological evolution. We gather around fire to warm up as well as to bond, socialize, learn, imagine, and cook our food. We use fire to make tools, thereby improving our lifestyle and increasing even further our chances for survival. No doubt humans decreased their chances for survival by using some tools and/or fire itself as weapons against each other. Much, much later, by about six thousand years ago, fire was used to melt hard copper alloys and mold them into refined tools. Today every technological achievement can be connected to fire—fuel is somewhere burning, creating electricity and powering our many devices, which to be made in the first place require fuel as well. Fire has been culturally so impressive and revolutionary that it became part of many rich ancient mythologies. For instance, in ancient Greek mythology the titan Prometheus defied the orders of the Olympian gods, stole the fire that was their exclusive secret, and brought it to humankind (more about this myth in chapter 4, "The Mythological Era"). Like Prometheus, scientists today aspire to "steal" (or discover) the other secrets of nature. Fire has also had a significant role in our scientific theories of nature (see, for example, chapter 12, "Heraclitus and Change" or chapter 15, "Empedocles and Elements").

It was the light of fire that dispelled the darkness of a cave, warmed the body and soul of our cave ancestors, and inspired them to paint their early wall paintings. If we accept that necessity (e.g., the instinct to survive) precedes luxury (e.g., of painting), then lighting fire at will might have been discovered by *Homo sapiens* before cave painting, or at least around the same time. For, as a result of the necessity to survive, it is possible that our ancestors foresaw the numerous benefits of fire and thought long and hard on how they could control and implement it in their daily lives.

Primitive Religion

Homo neanderthalensis (commonly, Neanderthals) might have been the first species who buried their dead. *Homo sapiens* practiced simple burial probably as early one hundred thousand years ago. But there is some evidence that their inauguration of painting coincided with their practice of more elaborate burials, which included burying of various grave artifacts (such as stone tools, animal parts, food) that might have been viewed as significant in the life of the deceased and quite possibly might have also been considered significant in an afterlife. Sculpture was another form of art that was enjoyed around the period of painting. Findings include various types of figurines, such as the "Venus": women with exaggerated bellies, buttocks, and breasts signifying perhaps fertility or other beliefs, carved from mammoth ivory or stone. Burials, painting, and sculpture are activities arguably indicative to a sort of primitive religious outlook, even though organized and systematic religions began with urbanization and the consequent development of specialized professions such as priests or group leaders. In the subsection "King-Priests, God-Kings, and Theocratic States" (of the next section), we will see how a priest and a group leader often were the same person, the king-priest.

Hunting

The hunting techniques eventually improved because in addition to spear throwing, which facilitated killing large dangerous prey from a distance without the risky physical contact, there was also the invention of the bow and arrow. Cave art depictions suggest that the bow and arrow had been in use by about twenty thousand years ago. But our evolutionary cousins, the Neanderthals (with whom we share a direct common genetic ancestor), did not invent such weapons. This, as we will see in chapter 6, "The Birth of Science," might have been one from a combination of reasons that caused their extinction between thirty thousand to twenty-five thousand years ago. *Homo floresiensis*, the most recently discovered type of human, emerged ninety-five thousand years ago in the island of Flores in Indonesia, stood about a meter tall, had a small brain size, but finally it, too, became extinct less than twenty thousand years ago (although its status as a separate *Homo* species is still controversial).[9] Thereafter *Homo sapiens* remained the sole survivor from the once diverse human family tree. But to be able to advance the species and better their quality of life they had to be inventive. Their first great idea directly involved the life of other species.

Herding

On the road to civilization a critical step toward a settled life was achieved through domesticating animals, an idea that soon led to herding. This happened about ten thousand years ago, at the end of the glacial period. The two important consequences of herding were reaching a surplus in available food and evolving toward a seminomadic way of life. The latter was a transitional step between the lifestyle of the once strictly nomadic hunter-gatherer and that of a fully settled farmer.

Herding and agriculture evolved about the same time, approximately ten thousand years ago, and their implementation was a turning point in the development of human culture. I believe herding most probably preceded farming, since herding would be a natural outgrowth of the hunting and gathering way of life, whereas farming is the result of a longer, tougher, and more organized type of work, which requires a far more thoughtful state of mind, since one must recognize the value of a crop and the hard work it takes to till the soil for months to achieve a harvest that can sustain oneself, one's family, and one's community. Thus if we assume things progress gradually, then the spontaneous nomad hunter-gatherer would naturally evolve first into the semi-spontaneous seminomad shepherd and then, eventually, the settled farmer.

Herding might have evolved by realizing that by capturing a herd of animals and keeping it alive (so the potential meat does not spoil) the daily meat could be secured for weeks, possibly months on end, making it much easier to survive without the uncertainty of daily hunting. Group members could kill an animal or two per day from the captured herd as their need required, but by the time they killed all captured animals of a herd, new ones were being born. So as long as people cared for the animals of the herd (e.g., fed and watered them, aided their reproduction), the supply of meat, milk, fur, hides, bones, and the like could be limitless.

Moreover, herding not only secured the daily meat on an ongoing basis but also provided extra resources such as new types of food (cheese, butter) and created the need for new occupations. Since not all animals of the herd had to be slaughtered at once, some were always available for milking. Some milk was consumed fresh and some was processed into butter and cheese. Animals could also be used as beasts of burden to aid humans. Ways are found to preserve the surplus food for the future. Preserving surpluses and planning for the future is a good thing and a sign of the human tendency to become urbanized. With a flock of animals under his control, a shepherd did not have to hunt every day to feed himself and his family. Thus as long as he maintained a healthy flock, food was plentiful for long periods of time. The first domesticated types of animals were goats, sheep, and cows. Animals of this kind do not eat meat, thus they do not have the natural drive to kill and are more peaceful compared to predators. Thus they were tamed easier. In addition, they eat what humans do not: grass and other types of plants, and besides meat they also give us milk. Goats, sheep, and cows were abundant in the Fertile Crescent and had been hunted before their domestication. The Fertile Crescent, which is the region from Egypt to Mesopotamia bounded by the Syrian Desert from the south and the highlands of Anatolia from the north, was among the earliest regions to develop a settled lifestyle and agriculture.

Semi-Settled Life

In addition to providing a food surplus, herding created the need for a seminomadic lifestyle. Group members no longer had to follow the migration of their prey. They could decide where their home would be, at least for an extended period of time. Animal domestication was therefore an important first step toward the settled life of community living. While a shepherd still had to be on the move in search of virgin grazing grounds, he was also required to have several temporary home bases (perhaps seasonal) where he could carry out various new activities: milk the animals, process the extra milk into butter and cheese, smoke and dry the extra meat in order to preserve it, care for the animals (e.g., in breeding), and develop specialized tools. Their semipermanent home bases might have been in places with rich grazing grounds such as fertile riverbanks.

The flocks were led to grazing fields during the day. But to limit the straying, by night they were brought back to a fenced base, thus a home. Fences were probably built by piling wood or rocks. Some seasonal dwellings might have existed even before herding, as long the natural resources of a region could sustain the nomads for some period of time. The benefits of seminomadic lifestyle might well have created the desire for an even more settled way of life and hence might have triggered the ingenuity in humans to search for ways to achieve it.

Idiosyncrasy

This transition from the nomadic to the seminomadic lifestyle was heralding a critical change in human behavior. Specifically, with herding, the once spontaneous hunter-gatherer who was basically searching for food whenever he was hungry gradually transformed into a shepherd, who, by domesticating animals, figured out a way to have easily accessible and abundant food for the future! Planning for the future cannot be a bad thing for civilization. But by far, this uniquely human approach to survival became even more refined, particularly with the domestication of plants, which led to farming and an even more settled lifestyle. The biology of the human brain was ready for it, for by this time the central part of that organ had evolved the ability to plan for future activities and wait for their resulting rewards to be fulfilled months and even years later.

FROM AGRICULTURE TO CIVILIZATION: 10,000–6,000 YEARS AGO

Agriculture

The decision to domesticate plants was one of the most consequential events in all of human history. This culturally explosive phenomenon, which occurred about ten thousand years ago, marked the end of the Paleolithic Age and caused the start of the Neolithic Age ("New Stone Age"), which lasted until about six thousand years ago. Plant domestication led to agriculture, which required a settled way of life, hence the development of urban living in communities that grew and in turn led to the rise of civilization.

This critical transition in lifestyle, from the nomadic hunting-gathering to farming villages, which gradually evolved to technologically advanced and densely populated cities with social and political structures in place, is known as the Neolithic Revolution. This was not some singular event; rather, it generally occurred independently at several locations worldwide during the Neolithic Age (though roughly simultaneously).

Although the rate of development differed from place to place, agriculture and urbanization soon became widespread. New ideas evolved from within specific communities, but ideas were also exchanged by people from different regions through their various contacts. Because of the need for fresh water, the first urbanized centers tended to develop near the fertile banks of permanently flowing great rivers. These were the Fertile Crescent (extending from Egypt by the Nile to Mesopotamia by the Tigris and Euphrates Rivers); south of Sahara; the Indus Valley; the Yellow River and Yangtze River valleys in northern and central China, respectively; and Central and South America. With an abundance of fresh water farmers did not have to depend on inconsistent rainfall to water their fields, their flocks, and themselves.

A factor contributing to this transition appears to have been significant climatic changes that not only made the weather in general more comfortable but also allowed for the evolution of abundant and edible new types of vegetation in certain areas of the planet. These areas later evolved into the first agricultural communities. Specifically, by twelve thousand years ago the ice of the glacial period began to melt. The climate was then becoming warmer and the landscape changing considerably. By ten thousand years ago the glacial period ended, and thereafter the climate has generally stabilized into the familiar one we know today. By about the same time huge quantities of wild wheat were growing in the Middle East (the region between the eastern Mediterranean, the Persian Gulf, and the Caspian Sea). Wheat was the first domesticated plant, followed by barley.

Through their adventures in the wilderness the hunter-gatherers must have observed so many times how seeds that fall on the ground from plants, in time, grow to become the same type of plants as those they had fallen from. At the right time, such observation must have inspired the deliberate planting and watering of seeds and in general the systematic cultivation of land. But the right time occurred when people came to first appreciate the benefits of food surpluses. So while they were still only hunter-gatherers or shepherds, semi-settled seasonally, before deliberate cultivation had begun, they might have taken advantage of naturally occurring surpluses of food (e.g., wheat). They harvested and preserved these extras for consumption in later days or months. Because the surplus of food was used so beneficially, it occurred to people to control and enhance food production further by deliberately planting extra seeds and aiding their growth. When harvest season came and abundance was everywhere, people became convinced that agriculture was a vital alternative to hunting-gathering or herding alone. Of course these groups continued to hunt and keep flocks. But since plants added to the animal food source and created even more of a surplus within a short time, agriculture would be a major occupation. Gradually, during the Neolithic Age the nomad hunter-gatherer transformed into a settled farmer, and the idea of systematic human-made cultivation of the land had been fully established. Everything that pertains to the notion of civilization is a consequence of agriculture!

Settled Life

Agriculture ushered in a more fully settled lifestyle. Thus with it we have the inauguration of urbanization. The farmer adapted to a domestic lifestyle by living permanently in settlements close to his fields. To put food on the table each day, the farmer had to commit to and endure hard work and careful planning for several months: clean the fields from unwanted plant growth, plow the soil, find seeds (e.g., of wheat, barley, peas, lentils, rice, beans, avocados, potatoes, squash, dates), plant them, redirect water from a nearby river to irrigate the growing crops, harvest them, eat some fresh and process some of the harvest into new types of food (grind grains to make flour, moisten it, then dry it with the heat of fire and make a nutritious bread), preserve and even trade the surplus, save seeds for next year, rotate the crops, keep his beasts of burden healthy (horses, donkeys, mules, oxen, camels), attend to his flocks, make/fix tools, and so on. But this hard work was rewarding. Agriculture increased food productivity, created and exploited surpluses, raised the confidence of the people, and in general brought them prosperity and especially security, a feeling we all understand even today once the daily food of future months is secured way ahead of time. Furthermore, since successful agriculture was requiring the collaborative effort of many people (e.g., in harvesting the crops or constructing specialized tools) the benefits of coexistence were being appreciated and pursued. The settled lifestyle of the farmer promoted human collaborations but also conflicts. Conflicts could be avoided and collaborations could be enhanced if farmers chose common leaders who would develop a set of rules fair to all. This gradually led to organized communities called city-states.

From a Village to a City-State

The first villages were simple and extended for just a few acres. They had anywhere from twenty to fifty square or oval mud-brick multiroom houses. The village population was perhaps a hundred to a few hundred people, most probably close and distant relatives. Rooms were used for sleeping, cooking, and socializing but also to store tools or the surpluses of food. The daily life in the village was basically agricultural and pastoral, thus all villagers had similar skills. Family and perhaps village leaders would have been the elders since these were the people with the most experiences.

With time, about six thousand to five thousand years ago, both the village and human culture underwent an extraordinary transformation. The village grew to become a city-state: a complex, self-governing, organized community of a few thousand (thus no longer made of just blood relatives), that included a city and its surrounding farmland and grazing pastures. Life in a city-state created complex human interactions. It promoted innovative ideas and rapidly accelerated their dissemination, and it set the stage for the rise of human civilization. With time the city-state would become more systematic, more advanced, safer, and its people more sophisticated and with a better understanding of the world around them. The settled agricultural economy provided new challenges but also opportunities and encouraged the flourishing of numerous specialized professions. In fact the successful operation of the city relied on the continuous emergence of innovative, specialized pursuits, as is the case today. It also necessitated that people collaborate on just about everything. One depended on the skills of another. The farmer depended on the craftsman for tools just as the craftsman depended on the farmer for bread and milk.

As a result of a predictable food supply and the safety that communal living provided, urbanization spawned an increase in population. Labor-intensive agriculture made it practical for a farming family to have many children (a common practice even today in agricultural villages). When the boundaries of city-states began overlapping, the cities could decide to unite into one nation or fight with each other. The city-states by the Nile followed the first path and so by about 3100 BCE, Egypt was the first nation of the world. Later the many city-states in the Yellow River Valley became one unified and centralized Chinese Empire around 221 BCE. On the other hand, the Mesopotamian city-states and later the Greek city-states generally tended to follow the second path.

The inhabitants of the first villages must have developed a feeling of group identity, which evolved into what we today call nationality as the settlement size increased from a village to a city and then to a nation. Now, what kind of a ruler and political system did a city-state have and why?

King-Priests, God-Kings, and Theocratic States

Before we urbanized ourselves in communities large and small the law of the people was to be found in their customs. In growing urban environments customs became increasingly more complex, thus peaceful coexistence in a city and its effective operation required a common vision and a leader. Such vision was often provided by religion, which itself became a custom, for organized religion (as we will elaborate in chapter 4, "The Mythological Era") was born with the onset of structured communities. Consequently priesthood was among the first type of specialized professions everywhere. But so was the profession of village or city leader. And because religion and politics were evolving together to accommodate the changing cultural, religious, and political needs of the city-state, the first states were theocratic. This means that the ruler of the people was both a religious leader (a priest) but also a political one (a king)—that is, there was no separation between church and state; religion and politics had from the beginning been closely intertwined. Thus the institution of king-priest and of a hierarchical society (a stratified society in which various priests would have different amounts of power in the governance of a city) developed almost simultaneously with the first city-states. But why?

Priests were the first type of people who contemplated the world in an abstract way; they were looking for what they viewed as hidden meaning and purpose in nature that needed to be unearthed and explored, hence to the other members of the community they appeared to demonstrate certain wisdom. As a result of the insights they voiced, many in the community regarded them as among the most knowledgeable of people. Now, as will be argued in chapter 4, the earliest religions were cults focused on natural phenomena that influenced the crops people grew, so a priest had to master the right rituals in order to encourage the spirits believed to be in nature to produce rich yields. Consequently, priests were socially significant because of their efforts to secure people's daily food, their paramount concern in the struggle of life. But these priests were also politically significant because they were perceived to be the ones most connected with the honoring of powerful gods or spirits. This is so because priests, as specialized religious professionals, were those who devised and implemented the various religious rituals of daily life in the village or city, and were thus considered *inspired*, *literally*, as if a divine spirit entered their body during a ritual, advised and generally endowed them with valuable knowledge. Why this was thought so will be discussed in the section titled "From Dreams to Spirits" in chapter 4. People imagined, for example, that during the rituals, the laws of the city and of people were given to the priests *directly* by the gods (through divine revelations or inspirations)—in fact it is in this sense that the political system was theocratic, "rule of god" in Greek. Hence, not only priests were regarded as divinely endorsed (thereby creating a link between ordinary people and their gods), but so were city laws (including the moral code of the community). Since both city law and priesthood represented the will of the gods, both earned immeasurable authority. And so priesthood status was elevated to prestigious, mighty king-priest status: a priest was in charge of both the religious as well as the secular/public matters of a city-state, thus he was also a king. There were in fact instances (for example, in Egypt) in which a king-priest himself was regarded as a god by his people, thus a god-king.

Temple offerings by the people—anything from food and personal items to flocks of sheep, land, even people—that were meant for the gods were in actuality owned and managed by the ruling priests, making them wealthy and powerful. And to preserve the special family privileges, kingship naturally became hereditary (or in general restricted to the socially privileged). All political, military, religious, and economic powers were centralized and controlled by the king's entourage and ultimately the king himself, who answered to no one. Democracy, which was born in ancient Greece (and played a significant role in the birth and development of science), had not been an option for the first ten thousand years of civilization (see next section). A good king-priest was respected for promoting peaceful coexistence and organizing the effective functioning of his own city. This was a challenging task, especially during early human history when rules and regulations were still in their infancy and thousands of unrelated people were living in relatively new social settings: living in harmony with one another in the close spaces of a large village, town, or city often meant that some of one's freedom had to be sacrificed for the greater good and stability of the community. A common religious vision probably eased such problems. At the same time, however, while a common vision and belief in a shared pantheon of gods or spirits could unify one city, it could also cause conflict between cities with different visions and pantheons, an age-old challenge that persists even today. Still, the benefits of urban life were immeasurable whether they were cultural, technological, political, religious, artistic, economic, or of some other form. But the single most important consequence of urbanization was the creation of free time!

Leisure

Food surpluses and specialized professions meant that at least some people (usually the more well-off) had certain leisure time to devote to anything their heart desired, including the arts, one-to-one personal and warm socialization with fellow humans, even contemplating abstract matters that at one time were the province only of the priests. Religion and the priesthood had been the first of these. With time philosophy, literature, history, science, and mathematics followed (though after the era covered in this section), as well as many other things that the notion of culture entails. All of these actual and potential benefits of communal living have been the joyous pursuit of all types of peoples all over the earth as a result of urbanization. Settled life fostered the evolution of the human intellect. The improvement of spoken languages as well as the invention of writing were the results of the evolving needs of an urban existence. They accommodated more efficient communication for trade and commerce, tool refinement, and techniques of farming.

FROM WRITING TO THE BIRTH OF SCIENCE: 6,000–2,600 YEARS AGO

The discovery of copper probably occurred as early as 6500 BCE, but its use was becoming widespread by about six thousand years ago, causing the transition from the Neolithic to the Chalcolithic Age ("Copper Age"). With the use of copper, tools improved significantly. The island of "Cyprus gave its name to copper"[10] because it was rich in it and among its earliest sources. Copper and tin form a hard alloy called bronze, which revolutionized not only everyday tools but also warfare. About the thirteenth or twelfth century BCE, the Greek and Trojan heroes fought with bronze weapons (Homer's *Iliad*). Nonetheless, bronze was no match for tougher iron. Formed in the cores of supermassive supergiant stars toward the end of their life before they become supernovae, iron makes stronger-edged tools. The Iron Age began around 1200 BCE. The combination of fire and iron has ever since been shaping both our technology and the general direction of our civilization.

Sumerians were probably the first to develop writing sometime in the fourth millennium BCE. Egyptians followed soon thereafter (if not simultaneously or prior). Writing is a momentous invention that accelerates the rate of progress because recorded information helps future generations to learn faster, more accurately, and more systematically all the accumulated wisdom of their ancestors. The need to keep track of the quantities of items accumulated and traded might have necessitated the invention of writing, including primitive forms of mathematics. With writing we have the beginning of recorded history and the accounting of what went on before (prehistory). The critical role of language (both in its oral and written forms) in the intellectual evolution of humankind (in fact even in our physical survival) will be elaborated in detail in chapter 6, "The Birth of Science." We will see, for example, how significant the Greek language was in the evolution of ideas and the birth of science.

Egyptians were obsessed with the afterlife, and around 2686 BCE they built their first of several elaborate stone burial tombs, the monumental pyramids. At the end of the third millennium BCE we find the first inhabitants of Greece. They were obsessed with life—Hades or the underworld for them was a gloomy place. Phoenicians invented the first type of written alphabet around 1050 BCE. And by 776 BCE the first Olympic Games were held. The social and political reforms implemented in Athens in 594 BCE by the Athenian leader Solon were the first crucial steps on the road to a democratic system of governance. In fact, the sixth century BCE was intellectually explosive globally.

Since sacredness is not easily challenged, theocracy proved generally repressive to religious or other innovations. So for millennia there was no significant evolution in religion. But sixth century BCE proved to be astonishingly different. The world at that time saw an outburst of religious and secular worldview reforms. This was seen in China with Lao-tzu (Taoism) and K'ung Fu-tzu (Confucius); in India with Buddha; in Persia with Zoroaster; in Israel with the Jewish prophets; and in various Greek city-states, too. Greece was a haven for the mystery religions whose subtle influence on the birth of science is explored in chapter 5, "Religion and Science." In southern Italy we find Pythagoras, whose school combined religious, mathematical, and scientific studies, and in Ionia (the Hellenic region of Asia Minor with the islands nearby) emerged the first natural philosophers whose rationalistic approach about nature marked the transition from mythology to science. Among several factors (to be introduced in chapter 6, "The Birth of Science") that contributed to the various reforms in ancient Greece was that Greek city-states (those that began forming after the fall of the Mycenaean civilization, around 1200 BCE) were not theocratic—"Greek gods do not give laws."[11] After thousands of years of no substantial progress in religion, any religious transition in itself, such as those that took place in sixth century BCE, was a promising prospect.

This chapter's brief human history finishes with the total solar eclipse of May 28, 585 BCE, foretold by Thales of Miletus in Asia Minor. We know the exact date of this phenomenon by using the modern calendar that predicts eclipses to calculate those that occurred in the past. The eclipse occurred the day of a great battle between the Lydians and the Medes. It suddenly brought a semblance of night to day and shocked soldiers and kings from both fighting armies. They interpreted the natural phenomenon as a divine omen against their continued hostilities. Frightened by the possibility of having angered their gods, they agreed to end both the battle and their six-year war. Thales was the first to use purely natural causes to explain the phenomena of nature. Others before him explained them supernaturally through myths and superstition. Since Thales flourished around the time of the famous eclipse, that era marks the birth of science. Parenthetically and interestingly, whether or not the eclipse was a divine sign is arguable among believers. But what is equally interesting is that this or any other eclipse also has a natural cause and a rational explanation.

Religion, philosophy, and science were all consequences of urbanization, a phenomenon so culturally impressive that stories from when it was first implemented seem to have survived until today: the story of the Golden and Silver Ages can be traced back to those early days and so can the legend of the lost city of Atlantis.

THE GOLDEN AND SILVER AGES

Hunting is risky but exciting. Farming is safe but dull. To hunt you get up and go, you are spontaneous and carefree. You have adventures with dangerous, magnificent, and strange animals in the wilderness. You move to be near the game that you hunt. The stories you tell your kids are thrilling, and you are the hero of the family. The gratification of the daily kill is instant though uncertain: but the food of the day must be found every day. However, with farming you settle down, plan for months ahead, and commit to a life that finds you tilling the soil. You have the safety of a home and a community, but your daily stories are ordinary. The gratification of farming is gradual but secure: planting occurs during one season, harvesting during another, but a rich harvest guarantees food not just for a day but for months ahead! This lifestyle difference has been a subject of ancient oral folklore but also of ancient written tradition, such as Hesiod's poem *Works and Days*.

Hesiod is an ancient Greek poet but also a farmer who lived around the late eighth century BCE. Among other things, his *Works and Days* describes the five Ages of Men (the stages of humanity since its emergence). The first was the best, the so-called Golden Age, a prehistoric utopia that, according to the poem, appears to have been the preagricultural, pre-urbanized era of carefree wanderers, the hunter-gatherers who lived spontaneously day by day. Incidentally, it has been speculated that the story of the Garden of Eden is from the Golden Age.[12] The era of agriculture and of the farmer was the second, the so-called Silver Age. Farming was committed, hard, tedious work, and farmer Hesiod had firsthand experience of it. So after a long laborious day in the fields, from dawn to dusk, farmers would go home looking forward to a well-deserved plate of food, worrying however about the constant challenges of tomorrow (the weather, their crops, their animals, their tools), and during their restful hours they would, I am quite certain, recollect with nostalgic envy on the carefree, uncommitted lifestyle of their not-so-distant ancestors, the hunter-gatherers (or in general those who had never committed to farming). Calling the hunting-gathering era golden seems to have been the reminiscing words of a tired, worried, reflective farmer.

Interestingly however, for Hesiod only carefree hunter-gatherers, from the Golden Age, or shepherds in general, were good enough to be "loved by the blessed gods,"[13] whereas hardworking farmers (like himself) from the Silver Age were not loved because they were "less noble by far."[14] Perhaps the challenges of coexistence in an urbanized environment were too great for the settled farmers. And so "they could not keep from sinning and from wronging one another, nor would they serve the immortals, nor sacrifice on the holy altars of the blessed ones as it is right for men to do wherever they dwell. Then Zeus the son of Cronus was angry and put them away [annihilated them]."[15] Incidentally, in the Old Testament book of Genesis, Cain was a farmer, and his brother Abel, whom he killed, was a shepherd.[16] Now who were those people who were annihilated by Zeus? Where were they from?

THE LOST CITY OF ATLANTIS

Hesiod's sinners, from the Silver Age, might have been a general reference to city people, in particular those corrupted by the newly found temptations of urban lifestyle. Or they might have been a specific reference to the corrupted

Atlanteans, citizens of the lost city of Atlantis, who, by Plato's account (from his dialogues *Timaeus* and *Critias*), were defeated by the prehistoric Athenians with Zeus's help, and their legendary city was afterward sunk by the gods in a single day and night.

Plato places chronologically the destruction of Atlantis about nine thousand years before the time of the Athenian leader Solon, approximately 11,652 years ago, an era that roughly coincides with the onset of urbanization. He describes Atlanteans to have initially been people of nothing but virtue, only to later become of nothing but corruption. Since both groups of people, Hesiod's Silver generation and Plato's corrupted Atlanteans, were, according to each author, punished by the wrath of Zeus, and also since both groups are placed chronologically from around the era of the onset of agriculture and urbanization, then I think is plausible that they are one and the same group. If so, then Hesiod's "golden race of mortal men . . . loved by the blessed gods,"[17] the hunter-gatherers/shepherds of his Golden Age, might have been Plato's virtuous Atlanteans before their fully settled agricultural lifestyle or during its early transitional stages when things were still simpler and people's minds were purer. Hesiod's "silver [race] and less noble by far,"[18] of the farmers of his Silver Age might have been Plato's corrupted Atlanteans well into their urbanized lifestyle, both unable to deal effectively with the new, ever-changing, and demanding challenges of peaceful coexistence, such as increasing populations, limited resources, territorial disputes, land ownership, new codes of conduct, different overall philosophical outlooks on life, and ultimately driven to obliteration. Are we, the modern humans, managing better?

So among humankind's earliest cities, the most prosperous and successful might have been Atlantis, whose sudden violent destruction by possibly a natural cause (such as flooding or earthquake), and/or by human-made ones (such as greed, deceit, slavery, internal conflicts, external wars, and generally corruption), left such long-lasting and legendary images in the minds of those early and impressionable urbanites (either Atlanteans who survived, or others who visited the great city before its obliteration), that the tales they had told their children, exaggerated, have since then been traveling through the continuum of spacetime and stinging the imagination of all those who have been hearing them. If Atlantis is discovered to be a real city destroyed by war (and not just some imaginary city of an allegory), then Plato's account will be a written reference of the oldest human war, which ironically would coincide with the onset of civilization. It would mean that, as we were getting civilized we were also preparing for war, a still-persistent irony of civilization.

CONCLUSION

Agriculture led to an urbanized culture and eventually to developed civilization. In turn, the coexistence and interaction of the many groups nurtured the development of the human intellect and precipitated the pursuit of a multitude of innovative and specialized professions. Thus hunting and gathering were no longer the only occupations that could secure one's daily food. But by far the most revolutionary effect of civilization was the creation of leisure time! For free time created the potential for abstract thought and the development of sophisticated religion, philosophy, and ultimately science. But the scientific view of nature is the newest, only about 2,600 years old, whereas the mythological is the oldest, at least ten thousand years old—as old as civilization itself. How did we view nature before the advent of science? Why did the mythology develop? When was religion born and what triggered it?

CHAPTER 4

THE MYTHOLOGICAL ERA

INTRODUCTION

T he worldview of the first 7,400 years of civilization was purely mytho-logical. It was an intellectually unrefined era, nostalgically simplistic, and superstitiously phobic. Nature was imagined to be animated; natural phenomena were considered random, unpredictable, and the work of mysterious supernatural (not subject to any physical law) forces (e.g., capricious, anthropomorphic spirits or gods). There were nature deities (e.g., the sun-god and the sky-god), deified ancestors, plant and animal spirits, idolatry, totems (animals believed to have been the ancestors of tribes), king-priests, god-kings, and spirit-based political systems. Hence belief in the supernatural is literally as old as civilization—for what we today refer to as ancient mythology constituted humankind's earliest form of religion. While influencing each other since their birth ten thousand years ago, religion and civilization have been evolving from their crudest forms to their more sophisticated ones today.

DEFINITION OF RELIGION

Religion, at least the primitive kind, is the belief in the existence of beings with powers far greater than that of humans, with whom humankind wants to open communication and cultivate a relationship. All of the natural forces that humans did not control themselves were placed under the direct control and supervision of a spirit or god, which had to be understood and placated to ensure that these forces could be managed and made more predictable to the service of humans. With this in mind, the two culturally remarkable practices of Paleolithic humans, the possibly one-hundred-thousand-year-old practice of burying one's dead and the thirty-thousand-year-old practice of cave painting, are activities that while their mere practice in itself does not constitute religion, it may be a vague indication of an early tendency for religion.

BIRTH AND ENDURANCE OF RELIGION

Religion was born after humankind first concluded that such beings with higher powers—what came to be called gods—existed, and specifically at the moment humans felt the need to establish and nurture a relationship with them. This birth is generally believed to have occurred about ten thousand years ago with the onset of urbanization. The factors that caused this birth and the nature of the first gods are introduced later. What caused religion to endure, evolve, and become an essential part of human culture was when the idea of a human-god relationship developed over time into a great human need and thus a lifelong habit, passing systematically down through the generations. Specifically, this need triggered the development of various tribal rites and rituals through which people believed they could open communication with their gods, thereby developing, preserving, strengthening, and renewing it. As time went by, the rituals became more complex and constituted a significant part of communal life. This established and sustained the religious outlook as a way of life.

THE CHARACTER OF EARLY RELIGION

The goal of the religious outlook was twofold, initially selfless but later egoistical. It was selfless in the sense that through their rituals our ancestors aimed partly to secure the welfare of the tribe (and, consequently, of humankind in its struggle for life) by appeasing, befriending, encouraging, persuading, even trying to control and manipulate the gods to care for the people: to tend to the tribe's practical needs regarding food, shelter, health, and fertility. The character of early religion was tribal[1] not personal—in a sense, just as the search for food or shelter was a tribal activity or, to say the least, a family effort. It aimed to guarantee tribal well-being. Religiously speaking, initially the individual cared more for the welfare of his community than for himself! Little by little, when religion's character evolved to be personal (egoistical) as well (see the section under "Idolatry"), the gods were expected to tend even to the more abstract and personal needs of the individual, such as those regarding personal grief, happiness, knowledge, comfort, and hope, as well as meaning or purpose for this life and even in the afterlife. Humans worshipped and idolized the gods, built them shrines, and brought them offerings in the hope of receiving favors in return. "By gifts are the gods persuaded, by gifts [are the] great kings" says an ancient Greek epigram.[2]

But the goal of the religious outlook was also selfless on another level, for the rituals were not always aimed at securing what individuals (or the tribe) wanted from the gods, but also at what they could secure for the gods. Our ancestors literally sympathized with them in what they perceived to be their own divine struggles, for gods were imagined anthropomorphically—in our own image (with similar needs and challenges as humans). They felt or shared their gods' passions, their sufferings, their needs and challenges; they showed affection for them, expressed admiration and respect for them, even sacrificed for them (i.e., gave the gods something humans valued, such as food and drink, material valuables, or in extreme cases human sacrifice). (The act of sacrifice was indicative of the belief that not only did humans need the gods but also that the gods needed humans.) Some people devoted their lives to the gods (e.g., by becoming a priest in a temple in order to better serve the god who was thought to be literally living there). Finally, through the rituals, people wished merely and humbly to give thanks to these gods for their previous help (e.g., for making it rain or causing a rich harvest). Devotion to the gods quite possibly was even intended to promote harmony in the coexistence of humans and these divine entities.

CAUSES OF THE BIRTH OF RELIGION

While it is generally accepted that organized religion (with gods, their temples, priests, and rituals) was born with the advent of urbanization, more speculative is what might have sparked its birth and aided its development. No doubt a variety of factors might have played a role. These form three general groups: (1) theological, (2) biological, and (3) cultural.

The Theological Factor

Is religion (and, in particular, the belief in the existence of a god or gods) knowledge that came about via divine revelation, is it a human discovery, or is it a human invention that humankind created? If it is a revelation or discovery, then a god (or gods) is presumed to exist. If it is an invention, then a god (or gods) cannot necessarily be presumed to not exist, for arguably any human invention/creation might be, directly or indirectly, caused by a god. This is so because, while science can determine various properties of the universe causally, it cannot determine its first cause—what caused the universe (recall the section titled "Causality" in chapter 2 "What Is Science?"). In science there will always exist an initial unexplained axiom (a primary cause)—and as we will emphasize in the section "Unborn and Imperishable" in chapter 13, "Parmenides and Oneness," there is not, and there can never be, any scientific explanation of how something (e.g., the universe) can come to be from absolute nothingness. Hence the *why* of this first cause, whether it might be some god or related to some god (and thus, why the universe is what it is and why it exists), will always remain a subjective matter, and consequently so will the belief in the existence of a god and the cause of the birth and development of religion. A god of *absolute* powers (e.g., wisdom) is unprovable, for the only way to recognize, understand, and prove the absolute is if we ourselves had the same absolute powers. But we do not. So even if an absolute god were to reveal itself, the only thing we could logically be certain of is that the being revealed has great powers but not necessarily absolute ones—hence we would not know if it were the true god or just another being with powers merely greater than ours. The most central feature of religion, therefore, is faith, not causal knowledge. But if religion is a divine revelation, why shouldn't everything else that we come to know, such as science, be a divine revelation also? And if we were to accept that things are divinely revealed to us, then a question begs for an answer: what is the role of our mind?

Having said that (and leaving the nature of the primary cause of the universe to be a subjective issue), if we are still interested in pursuing a scientific understanding of the universe, we must remain of the conviction that the universe is inherently rational, that it obeys understandable causal laws (something proven repeatedly by science). And so everything that happens in it, including the birth and development of religion, has a rational explanation. Human biology and culture are part of a rational explanation concerning the birth and development of religion.

The Biological Factor

(a) A Biologically Evolved Brain but an Ignorant Mind

A requirement for critical inquiry of nature is a biologically evolved brain. Now, while a critical examination of the phenomena of nature has always been an impossibility for all other species—since their brains have not yet evolved to that crucial level required for abstract thought—for humans, it has not. By about ten thousand years ago, our prehistoric early Neolithic ancestors already had a brain that was as biologically evolved as our own. This early species was therefore as capable of critical analysis as we are now. And so for the first time they became profoundly curious about their surroundings and developed the desire to know the nature of the world they lived in. But without much prior knowledge as a point of reference on how natural phenomena worked, and without an advanced language to express themselves clearly, these first explanations were childish, irrational, dogmatic, mystical, and overly simplistic. For, while Neolithic humans were intelligent, they were like young children who have the biologically advanced organ of knowledge—the brain—yet an ignorant mind. Hence, naturally, early Neolithic humans' raw intelligence led them to model natural phenomena after the only thing they knew best: themselves. In their own image they modeled these phenomena with feelings, needs, motives, challenges, desires, pleasures, passions, and powers that were in fact not only greater than their own, but justifiably they were also imagined to be supernatural in their extension and capabilities because nature is impressively powerful (e.g., the thunder is unpredictable and loud, the lightning is fast and bright, and the sky is huge and ever-present).

Initially it was, for example, the *object* sun that was imagined to be animated and a god—the sun-god. That is, there was no reason to see an object, such as the sun, and think that it is something different (e.g., an anthropomorphic being) than what it appears to be. But with time, the attributes of the animated phenomena were imagined so much to be like and possess the attributes of humans (and/or of other animals), that the phenomena became tangible and anthropomorphic (or, in general, zoomorphic), and so in addition to humanlike behavior, they also acquired humanlike form. Consequently, object and god were gradually separated: for example, the object sun-god became the anthropomorphic sun-god, who, among other things, could control the rising and setting of the object sun. As in human culture, the gods were imagined to have specialized professions. The moon, planets, and stellar constellations of the zodiac are additional examples of objects that initially were object-gods that gradually evolved in the human mind to become gods of a zoomorphic or anthropomorphic nature (or a mixture of the two). In ancient Greece the father god of all, the heaven-god Uranus ("heaven," in Greek), and the mother goddess of all, the earth-goddess Gaia ("earth," in Greek) had similar evolutions. Uranus and Gaia initially were the physical sky and physical earth, but with time they were imagined as two anthropomorphic gods. Their son Cronus had a son of his own, Zeus. Uranus and Gaia gradually decayed to lesser gods, and mere grandson Zeus evolved to become "Father of men and gods."[3] Even more impressively, as Greek religion evolved from Homer's and Hesiod's Olympian to the mystery religions (the Eleusinian, Dionysian, and especially the Orphic), so was the status of Zeus attaining a sole and absolute divine supremacy as an all-encompassing father of all. So a deity who once was the natural object/phenomenon itself was gradually becoming increasingly disconnected from the object ultimately acquiring a divine existence of its own with new roles and attributes. Now, since capricious supernatural gods controlled nature and since humans could not know their divine minds, early humans thought that nature was random, unpredictable, and at the absolute mercy of these divinities. But humans hoped that their rituals could persuade the gods to act favorably toward them, and, so in a way, indirectly, humans hoped they could have a certain amount of control over nature.

Motion and sound, which are among the important characteristics of a living person (although not just of a living person), were also characteristics of the phenomena of nature; trees shake and appear to cause the wind (although what happens is the reverse; the wind shakes the trees), the sun and moon rise and set, the thunder is loud, the rain comes down from the sky. And so for our Neolithic ancestors various things and phenomena appeared as alive as they themselves, only more powerful. Their deification and personification followed naturally thereafter.

Because humankind recognized the benefits of communicating and culti- vating a relationship with the animated superpowerful phenomena of nature, to achieve it rituals were devised and implemented. The phenomena became instantiated as worshipped gods, and the stories about them emerged as myths. Some myths were only about gods, others about humans and gods. By one account of ancient Greek mythology, for example, Orion was once a handsome, skillful hunter who hunted with Artemis, goddess of the hunt.[4] But he was killed by a scorpion sent by Mother Earth because he was arrogant and threatened to kill all the animals in order to impress Artemis. At the request of Artemis, Orion was placed by Zeus among the stars in order to be remembered. Orion therefore was to the ancient Greeks the personification of a particular grouping of stars. To modern astronomers this grouping is *Orion the Hunter*, one of the most recognizable constellations from a total of eighty-eight that divide the sky—like the fifty states divide the country of the United States. In the evening it's easily spotted high in the sky and toward the south during the winters of the Northern Hemisphere (and visible during the summers of the Southern Hemi- sphere by looking north).

At first the myths were primitive, for although the human brain had the biological potential for progress, the mind was still imprisoned in absolute darkness. While awake, humans regarded nature as animated, zoomorphic, anthropomorphic, powerful, supernatural, and random; while asleep, it was seen as enigmatic, spirit-filled (the primitive interpretation of dreams, as will be seen in section "From Dreams to Spirits"), and often frightening. As civili- zation gradually evolved, its new challenges and needs made its myths more diverse, imaginative, and rich, and a number of these myths became the religion of some. So what we today refer to as ancient mythology constituted human- kind's earliest form of religion. Note that to the ancients a myth was not nec- essarily a fairy tale as the word "myth" has evolved to often imply nowadays, because many ancient myths have with time proven to be just that, fairy tales.

The first myths and attributes of the gods arose through early human attempts to interpret the phenomena of nature. The loudness of thunder and the brightness of lightning might have, for example, been associated with the wrath of the sky-god, and the wind with this god's breath (the latter, for example, was a belief of some North American Indians of relatively recent times). And so religion was humanity's first serious attempt to understand how the phenomena of nature work. Interestingly, therefore, our curiosity and desire to make sense of nature were among the stimuli for the birth and development of both religion and science. Nonetheless, the application of causality in each approach is fundamentally different: in religion the cause of the phenomena was divine (supernatural), in science it is naturalistic.

(b) Long Parental Care

Religion might have also been born as a consequence of yet another reason of biological origin, namely, the need to be cared for and the need to care for. The human species has evolved biologically in such a way that human babies lack the ability of self-reliance, so their self-preservation and general survival depend on the unconditional and uncommon long period of parental care. Ever since their birth and for several years thereafter, human babies need their parents' care and love. Without such long parental care, the babies would die. Now, if our physical attributes (e.g., brown or blue eyes) have a biological origin and explanation, why not our emotional ones (e.g., the need to be cared for)? If yes, then the biological need of the human baby to be cared for might exist in our genes and endure (subconsciously) throughout adulthood, driving humans to discover/invent their god through which they can continue to be cared for and to be loved. And so as we are growing older we are also naturally searching to supplant the gradually lesser parental care with a gradually greater divine care in order to fulfill the genetic need of being cared for. That is, in a similar manner that the genetic (biological) need to eat and drink (or to have sex) drives us to sources of food and drink (and to seek a sexual partner), the biological need to be cared for and to be loved (in fact even the need to care for and love, for parents have such need for their babies) might have driven us to seek God and religion. Through religion not only do we continue to be cared for (by the divine), but we continue to care for (the divine)—for, as we saw earlier, religious rituals have an egoistical and selfless objective.

Note that while the young of various other species have the biological need to be cared for, too, still such need is not as evolved as or as extensive as it is in humans. Even more importantly, unlike humans, all other species still lack the critical mind that could stimulate such biological need to more abstract endeavors (such as religion). My point is that we need to remember that the cause of something is often a complex interplay of several factors. Concerning the birth of religion the biological factor is coupled with a cultural one.

The Cultural Factor

With its evolving urbanized culture, humankind came to realize that its survival and overall well-being depended on nature's animated powers: the sun for light and warmth, the earth for sustaining plant and animal life (for food), the rain and rivers for water, the sea for fish. Humans could not control the rain or the light from the sun, but the rain-god and sun-god could. Intelligently our ancestors sought to establish communication with these natural powers and to cultivate a relationship in the hope of appeasing and befriending them, and through offerings to ask in return for their help in the human struggle to survive. So early humans deified and worshipped these natural phenomena. Modeling the human-god relationship after the human-human one was a consequence of the human culture: since individuals and groups befriended and formed alliances with other persons and groups for mutual benefit, by exchanging goods, a custom practiced for millennia before religion, that's how these humans thought they should behave toward their gods if humanity was to be successful in its relationship with them.

Fear of the phenomena (the personified powers of nature) was yet another culturally related reason that could lead to religion, for such fear could now be conquered through communication with these powers. Even the admiration, respect, or envy for such magnificent powers could lead to religion, for such feelings could trigger humanity's humbleness toward them, its desire to communicate with them in order to understand them better, perhaps even imitate them or sympathize with them in the hope of becoming like them. Nature was worshipped not only in terms of its particular objects (the sun, the stars, the rain, etc.) but also in terms of its natural processes, such as the cycle of moon's phases; day and night; the changing seasons; the birth, growth, decay, and rebirth of plants—for example, night was a goddess and so were the seasons. Humans' rituals were often an attempt to act out and imitate such processes and in general sympathize with nature (the gods), believing that through sympathetic magic life would imitate the ritual. And thus we humans could, at least in some aspects, be like the gods. *Enthusiasm*, for example, which etymologically meant that a god who was honored literally entered the body of a worshipper and influenced him in strange ways (e.g., spoke to him or made him move a certain way), was a state of mind hoped to be achieved by the worshipper in ancient Greece during the Mysteries (ancient Greek religious rituals of the historic era), in order that the person would feel godlike, not only physically but also intellectually—that is, he wished to know the mind of the god. The latter was such an ambitious goal that the desire to fulfill it, as will be discussed in chapter 5, "Religion and Science," might have been among the factors that sparked the rationalist viewpoint of nature and consequently the birth of science. For it had at some point been realized that achieving divine knowledge (knowledge supposedly exclusive only to the gods) ritualistically was really unattainable but could be attainable through one's rational analysis of nature.

There were as many gods as the vast array of phenomena in nature would permit—it was difficult to settle for only one supreme god just yet, and monotheism was millennia away. And since nature is ever changing, so were our human preferences for god. During a full moon for example, the moon-god is more important than the sun-god, but in the morning, the preference is reversed. Worshipping the natural powers as gods, coupled with the changing human culture, had gradually led to the belief that the phenomena of nature had even more attributes than initially imagined. It also caused human understanding of nature to be reversed, for with time, it was not nature's phenomena (the gods) who were imagined as taking human form; rather, humans began to imagine themselves as having the form of gods. In ancient times, for example, the mythical titan Prometheus created humans in the image of the gods by molding into human shapes a mixture of soil and water.[5] And, according to the book of Genesis, "God created man in his own image. In the image of God he created him."[6]

In general, as humans became socially more responsible and humane toward one another—that is, more conscious of their actions and the consequences of those actions—so did their gods and religion. Hence contributing to the development of religion was humanity's changing needs, challenges, and overall understanding of the world, all of which were coming about through the evolving complexity in lifestyle required by urbanization. The birth and development of science was yet another contribution to the evolution of the religious outlook, although that influence came much later as science was born about seven and a half millennia after urbanization and religion. Interestingly, however, unlike the birth of science, which occurred once and in one specific place (as will be seen in chapter 6, "The Birth of Science"), the birth of religion was universal.

UNIVERSALITY

Religion was generally born independently but universally (in all major urbanized centers of early civilization) and relatively simultaneously. It had no founder and evolved and became established through diverse custom and traditions. This universality is no surprise because the factors that caused the birth of religion were themselves universal: we all have about the same mental abilities to address our basic needs, challenges, and experiences, which themselves are generally also about the same for all, especially so in the early stages of civilization, because nature itself, which presents us with these challenges, obeys universal laws. The first types of gods were universal, as were the first types of religions.

FIRST GODS AND FIRST RELIGIONS

The prevailing view is that the first gods were those associated with the great powers of nature: things that were so powerful, admirable, impressive, fearful, wondrous, that appear so *alive*, such as the light-bringing bright, warm sun; the moon of many phases that illuminates the night; the strong whistling wind; the fire-carrying lightning; the vital rain; and the loud thunder. The concept of the divine might have extended to something huge and intangible, such as the ever-present thus immortal sky, which is strikingly so high above all, appearing to blanket and look over all, that from it the rain falls. Or something semi-tangible but still immensely powerful, also ever present and seemingly alive, such as the earth from which everything is born, nourished, grows, and on which humanity itself so much depends for food, shelter, and care. For this life-giving property earth was female, thus the earth-*goddess* was the mother of all life, and the sky, who rained onto her the fertile rain, was male, thus sky-*god* was the father of all, a religious notion common to almost every primitive religion. With this in mind, the first religions were cults aiming to warrant fertility. This is a reasonable development because people's main concern has always been their daily food, which after urbanization was coming mostly from farming, and so naturally people wished to have their priests, through correct rituals, properly encourage the gods to produce rich harvests. Furthermore, since these great powers of nature are themselves universal (experienced by all people in all lands alike) and timeless, likewise the first gods were the same universally as well as usually immortal.

But indigenous gods differed from place to place for various groups also made a god out of something minor and local to them like a stone, a fountain, a well, a river, a tree, animal fur, even a dead ancestor, by attributing to them supernatural powers that they did not in the first place appear to possess. Now, how did such minor and local things manage to become gods?

FROM DREAMS TO SPIRITS

The intellectually undeveloped humans of early civilization were in no position to understand the meaning of either death or dreams. But they had the mental ability to wonder about them and to seek explanations. So when the dead appeared in one's dreams, moving and speaking, it was imagined as if they were somehow still alive, as if something from them, let us call it spirit (psyche or soul), endured despite the physical death and decay of the corporeal body, and had the supernatural ability to *enter* the body of those who were asleep (the etymology of the term "in-spire"). For something that could be seen, even in a dream, was thought to be there where it is seen, thus in the body of the sleeping person. The images of a dream were as real to the unrefined but intensely curious mind of Neolithic humans as the shadows were to the prisoners in the parable of the cave. Now, to have the ability of entering the body, the spirit was imagined to be like a shadow or a breath or air, incorporeal and form-changing. Air or a cloud appears to have such properties, for example, they change their form and seem to fit everywhere. The spirit could also then move through solid walls (for, how else could it enter the body of one who is asleep inside a house with closed doors and windows), squeeze through small openings, even appear instantly and spontaneously here or there (e.g., from the body of the sleeping person to the place dreamt of). It could bring a message to the one who is asleep, give advice, endow the sleeper with knowledge, or take the sleeper's own spirit someplace far away. After the sleeping person wakes up and is reassured (by others in the same room with him) that his body never left his bed, a simple way to explain how he had dreamt to be in a place far away was to assume that he, too, had a spirit. Now such superhuman, supernatural mystical abilities of the spirit of someone who died could naturally lead to his deification, especially when the individual was already influential and respectful during his life. The burial and subsequent memorial rituals for his sake were elaborate to begin with but also were glorified and mythologized further with the passing of generations and time. Since the spirit of an important dead human ancestor could be imagined as a god, his children were also gods, and perhaps their children, too. Consequently, the origin of the human race could be traced back to some mythical, deified first parents. And the glorified life stories about some of them, those that endured time and stung the human imagination the most, had gradually become the stories about great gods.

There was, for example, stormy weather once upon a time. And lightning started a fire. Foreseeing its benefits, a brave man retrieved a flaming stick and brought the gift of fire to his group. But an angered thundering, blazing sky-god punished the unsuspecting curious man by burning his hands, for the man had defied god and stole the exclusively divine fire. Still to his children he was a hero because the knowledge of fire changed their luck. And to their children he was a creator since they could trace their life back to him. Some of those children were called Greeks, and such a man was called by them Prometheus ("foreseer"); rightly then he was their favorite and most admired titan god, who, by one account, created man from water and mud.[7] As implied by Hesiod's *Works and Days*, Prometheus reigned during the fabled Golden Age, the period of first-generation humans when all—people and gods—lived once in harmony together. But his further gift of fire brought his own downfall and also ended the Golden Age for his precious creation; people began transgressing and thus were separated from their gods. Still, through rituals and ascetics a fallen soul could once again be purified and reunite the race of gods, the Orphic (a mystery religion) held (as we will see in chapter 5, "Religion and Science"). Imagining gods forming families and living initially with humans (or humans to be the children of gods) is not difficult, since some gods, it appears, had once themselves been merely humans who had been enviably admired in live heroic actions, had been zealously dreamt about in inspiring great dreams, and had been eagerly imitated in daily life.

Both Hesiod in his *Theogony* (on the origin of the gods) and Aeschylus (ca. 525/524—ca. 456/455 BCE) in his *Prometheus Bound* imagined for example that Zeus, the king-god of the Olympian gods, god of thunder and lightning, punished his cousin Prometheus for stealing fire from him and bringing it to humankind. Fire was supposed to be the exclusive knowledge of the gods. So he tied Prometheus to a rock on a high mountain, naked and helpless. By day a wild giant bird would devour his liver. But Prometheus's suffering could not end, for he was an immortal god who could not die. His liver would regenerate at night, only to be devoured by the bird again the next day, in a continuous, eternal, and painful cycle. But at the end, after thousands of years, Prometheus is liberated by Hercules. Prometheus's punishment captures the fear of inquiry (and of the gods in general). But his liberation captures the human hope, if not the desire, for godly knowledge (e.g., of fire, of search and discovery), and the admiration (even envy) for one who dared defy the gods and steal it at will. So while defiance of the divine for the sake of knowledge is horrifically punished, it is also admired and worthy of being pursued, the story implies, for ultimately Prometheus is redeemed.

FROM SPIRITS TO IDOLATRY

Of course we dream not only of the dead but also of the living and of animals and inanimate objects. Analogously, the spirit of an inanimate object—say, of a mountain or an impressively large rock, or a lion's teeth, or a bear's fur—could lead to the object's idolization. Idolatry (in Greek, the worship of an image or an object) is the worship of inanimate objects imagined to be occupied by spirits and to have mystical powers.

Spirits were envisioned to move into other bodies or objects and assume a different kind of form or life than what they came from—an idea that forms the basis of reincarnation. So a bear's or a lion's spirit could move into a man's body, enriching him with its unique animal characteristics such as strength, courage, even new knowledge. Humans grew to admire and respect various animals as rivals in the struggles for survival, and so the deification and worshipping of animal spirits were common. Idols such as stones, animal teeth or fur, bird feathers or claws, even human-made artifacts (e.g., statues such as Venus figurines and generally various images of gods) could be regarded as embodying a spirit, too and thus considered alive and worthy of religious reverence. All these constituted idolatry.

Hence an idol was worshipped because it was imagined to be occupied by a spirit—although often it was also worshipped solely for the sake of the object itself disconnected from a spirit, for with time the distinction between object and spirit might have been lost. For instance, a lion's fur might have been considered embodied (in-spired) by the spirit of a dead lion. Wearing it, a proud hunter might have believed, could empower him with the same strength and courage of the lion. So while the fur itself does not have any powers, the lion's spirit that might reside in it can bring it power. Hunters and warriors prayed to the inspired object, extolled its virtues, encouraged it to help them, and brought offerings to it for all previous times it came through. But should this idol-god fail, the worship might quickly turn to a hatred; the deity then could be threatened. Incidentally, a Greek saying "Even a saint needs a threat," which is used allegorically in the modern times, probably has its root in prehistoric practices of idolatry. The idol was threatened and beaten, with the belief it could be persuaded to serve man better in fulfilling his ambitious dreams. But if this idol-god did not comply, it was regarded as useless, soon forgotten, and replaced with a new one. When the inspired fur did not live up to the hunter's expectations, it seemed as if the spirit decided to abandon it, in which case the hunter could also decide to abandon the fur and search for another mystical idol.

Trees themselves were initially worshipped as more than mere plants.

With time, however, they were imagined as the abodes of spirits that possessed various powers such as commanding the wind or rain, even causing fertility in plants and animals. Present-day maypole festivals during May harvest, which celebrate the end of uncultivable winter and the beginning of farmable spring in the Northern Hemisphere through dancing around a tall wooden pole raised from the ground, arguably attest to tree worship in the distant past. Wells were also worshipped for themselves, but with time a fairy (a spirit) was often imagined to be living in them. Throwing precious articles as offerings and making a wish expecting that it will be granted by the fairy was a common belief in early local religions. The modern custom of making a wish before tossing coins in a fountain certainly has its origin in the worship of wells, lakes, and rivers.

Now there were times when humans worshipped things, such as various animals, simply because of some certain quality that was admired or desired. To be successful in hunting and to survive the endeavor hunters had to be acquainted with the skills of the animals being hunted: for example, the quickness of the hare, the wiliness of the fox, or the strength of buffalo. Hunters respected and often envied such skills, feelings that led them to deify animals or their spirits. What was worshipped, however, was the species as a whole and not the individual members, attributing perhaps the species' unique skills to its mythical first parents. Early humans often regarded themselves as having a certain kinship with an animal or viewed themselves as the creature's descendants. This is known as totemism. While the hypotheses that triggered totemism are several, possibilities include the realization that humans and animals possess similar skills (even physical attributes, such as eyes, legs, etc.) and face similar challenges in the daily struggle for survival. Thus by associating with an animal humans might have thought they would begin acquiring its desirable skills and manage their lives better. A sacred animal often was not eaten.

So humans and many other things in nature were thought to have a spirit. Hylozoism (the view that all matter is alive) was a characteristic of humankind's developing religious worldview. Because *inspired* objects (the idols) were of various types, including small and human made that could be carried easily around by individuals, they could become personal gods. Without a doubt this must have contributed to the birth of personal religion. Before that, religion was tribal. Prayers, rituals, and offerings were practiced by the tribe (and for the sake of the tribe) and not by the individual (nor for the satisfaction of his individual needs). Of course, another factor related to the evolution of personal religion might be merely the desire of humans to have the gods attend to their personal needs and/or ambitions: if gods could be appeased and befriended by the tribe and for the sake of the tribe, why not also by the individual and solely for his sake? This is not something unlikely to have crossed the Neolithic's mind. The development of personal religion had lessened the priesthood role in religion.

Dreams had undoubtedly played a significant role in the birth and development of religion—an idea that was speculated also by Democritus[8]—even in the explanation of death and of an afterlife. The body dies (*expires*, i.e., the opposite of *in-spire*) when its spirit decides to permanently abandon it, but the spirit lives on. Interestingly, the interpretation of dreams as divine revelations or inspirations still remains a common religious viewpoint. In general though, how were minor, local religions converted into major and powerful state or kingdom religions?

STATE RELIGION

As different tribes, villages, and cities united (either voluntarily or by force), the gods that dominated were those imposed by the victor in battle, by the strongest culture, and/or those with the most appealing stories/myths. Some gods faded away, others fused into one another, and new ones emerged. Gradually, with advancing civilization, humankind's unique regional experiences caused local religions to evolve into great and diverse state religions. For example, the most powerful city's main god becomes a kingdom's most important god, too. Because a kingdom is geographically large, the main god's principal shrine might be either in the capital city of the kingdom or at the center of each of its major cities. Such a god (much like the king of a kingdom) is therefore literally at a greater distance from the worshipper compared to local gods (or leaders) in a small village. Consequently, that god's (or a king's) tangible familiarity is reduced, but being the most important and powerful god from all the local ones (as is a king compared to local chiefs), that god's (or king's) admiration and adoration become deeper, more mystical and abstract, and so does religion as a notion.

For example, a city, with its many and diverse citizens who are no longer blood relatives, can be more successful if the sentiment of social responsibility begins to evolve. People do not help just the blood relatives in a city but one another, too, as fellow city dwellers. In the evolving complexity of an urbanized lifestyle citizens realize their codependency and the need for mutual help (encouragement, sympathy, altruism, and communal good over the individual), and so they actively pursue innovative ways to make it work, for only then can the city and thus themselves benefit and progress. Since humans were becoming socially more responsible, so were their anthropomorphic gods (who are modeled after them). And since city life promotes the collective good over the individual, a city god is imagined to be as unbiased to any one city group of people. Such a god is imagined to be more universal, as the god of all the people of the city or state. This type of thinking can be applied to large empires. Like the king of an empire who looks after all his subjects, so, too, is the empire's main god/s imagined to be. However, local gods are also worshipped, and each city has its own individual guardian god. But again, with evolving civilization, the new personal religious sentiment is now defined by the more powerful state religion for which the main god (or a smaller group of main gods) has a more universal (state) appeal. This type of universality was heralding a religious transition, from polytheism to monotheism.

POLYTHEISM AND MONOTHEISM

Since humankind modeled its gods after its own experiences and social changes, then, just as a group of people had a recognized leader, a tribal chief, or later a king, in time so, too, was the case for a group of deities. Thus the preferred god from a group of many had its status elevated, and the notion of a supreme, or father god, or mother goddess of all is born. When with time the lesser gods were forgotten or willingly abandoned, polytheism had in some cases been replaced by monotheism.

The first time this was attempted was long after the birth of religion, during the historic times when the Egyptian pharaoh Amenhotep IV (also known as Akhenaton), who ruled Egypt from 1379 to 1362 BCE, believed that there was only one god, the sun-god Aton. His view, however, did not become popular. The Egyptian priesthood and consequently their followers adhered to their age-old polytheistic views. In Judaic tradition monotheism is attributed to Abraham at an even earlier era, a few centuries before Amenhotep, but this reference is found only in the Bible.[9]

Nonetheless, for Ulrich von Wilamowitz-Moellendorff (1848–1931), a renowned scholar of his time on ancient Greek matters, he who "upheld the only real monotheism that has ever existed upon earth,"[10] was Xenophanes (ca. 570–ca. 475 BCE), a pre-Socratic philosopher (and natural scientist) who believed that "there is one god, among gods and men the greatest, not at all like mortals in body or in mind. He sees as a whole, thinks as a whole, and hears as a whole. But without toil he sets everything in motion, by the thought of his mind. And he always remains in the same place, not moving at all, nor is it fitting for him to change his position at different times."[11]

Xenophanes has been called the first philosopher of religion. How philosophy was born of religion in Greece around his time and how the ancient Greek religion proved catalytic for the birth of science, will be the topic of chapter 5, "Religion and Science." Xenophanes criticized both approaches to religion—the popular polytheistic view as well as the anthropomorphic one.

ANTHROPOMORPHISM

Xenophanes wrote: "Both Homer and Hesiod have attributed to the gods all those things that are shameful and a reproach among mankind: theft, adultery, and mutual deception."[12] "But mortals think that gods are begotten, and have the clothing, voice, and body of mortals. Now if cattle, horses or lions had hands and were able to draw with their hands and perform works like men, horses like horses and cattle like cattle would draw the forms of gods, and make their bodies just like the body each of them had. Africans [it is Ethiopians in the actual Greek text] say their gods are snub-nosed and black, Thracians blue-eyed and red-haired."[13] Actually Greek philosophers did not "eliminate divinity from the world. They preferred to depersonalize their gods."[14]

In Xenophanes's time science was in its infancy. It was a transitional period from the purely mythological and superstitious worldview to the rational and scientific one. Did the mythological worldview aid in any way the birth of science?

RELIGION AND THE BIRTH OF SCIENCE

Greeks were seafaring people, and their travels put them in contact with diverse traditions and mythological worldviews. But when one is exposed to a variety of conflicting and inconsistent mythological explanations of nature, one must decide which one from such explanations is right or whether all are wrong. For example, what god is in charge of rain, the Ethiopians', the Thracians', both, or neither? Greeks (starting with the pre-Socratics) thought none were and in general that all mythological explanations of nature were illogical, aesthetically unappealing, and plain wrong. Their next step was the search for a more universal, naturalistic, and causative worldview. They managed to do this through science by examining nature rationally.

CONCLUSION

Religion was born after humans had first concluded that the phenomena of nature were controlled by beings with higher powers, in other words, gods, and specifically at the moment that humans sought to open and cultivate a relationship with such gods. Religion became firmly established and turned into a way of life when the idea of a human–god relationship developed into a huge human need and thus a lifelong habit, and passed systematically down the generations through rituals and traditions. With this in mind, and emphasizing again that what we today refer to as ancient mythology constituted humans' earliest form of religion, how could science evolve then, from within an age-old, time-honored, sacred, often frightening mythological (antiscientific, antirational) establishment in which myths, superstition, and the supernatural were the dominant worldview for at least the first 7,400 years of civilization? Could religion have been a stimulus for the birth of science?

CHAPTER 5

RELIGION AND SCIENCE

INTRODUCTION

I n the history of the world, it was religion that came first (about ten thousand years ago with the onset of urban living), followed by science much later (about 2,600 years ago). These two fields of inquiry have always shared a very intimate connection; they have been inspired by the need to understand the phenomena of nature in terms of abstract thinking. Now if "science must begin with myths, and with the criticism of myths"[1] then from some general point of view religion—the ancient mythology—may be regarded as the first and most basic type of science that had gradually been forming in the mind of the intellectually evolving human species as a means to understand all the unfolding phenomena in nature. Ultimately true science was born, but it was given birth from within a well-established and time-honored religious outlook. And such challenge, though formidable (for sacredness is not easily questioned or opposed), was remarkably overcome. Since people's religion, especially its evolution, tells a lot about the way they think—their aspirations, endeavors, hopes, passions, needs, desires, daily life challenges—anyone wishing to understand the success of this transition must search for possible hidden scientific tendencies and signs that might have existed within what appeared to be a purely religious outlook but really was not.

THE GREEKS OF MYSTERIES

At least this (the presence of subtle scientific tendencies in a society where the main outlook was the religious/mythological) was the case in sixth century BCE Greece, the era of the birth of science. The popular religion then was no longer the simplistic, placid, and happy Olympian[2] of Homer and Hesiod, where death nonetheless was a terrifying end, but rather it was the emotionally moving, ritualistically rich, and intellectually intriguing mystery religions[3] (such as the Eleusinian, Dionysian, and Orphic, called so because details of the rites were kept secret), in which the afterlife was a hopeful beginning. The ancient Greeks were generally religiously oriented, a fact evident from their art (statues, vases, and wall paintings), diverse and imaginative myths, rich pantheon of deities, famous oracles, temples, as well as their religious ceremonies, especially those of the mystery religions. But their continuously evolving religious outlook contained subtle elements indicative of their forthcoming transition from mythology to science. What exactly were these elements?

On the one hand, in the Olympian religion, nature was the playground of capricious, often immoral gods (although moral ones as well), with people and nature completely at their mercy. Human knowledge, and even actions, were believed to be decidable and controllable by the gods. Zeus, for example, could choose to strike with a thunderbolt, Eros could cause someone to fall in love, Apollo could heal as well as bring on a plague, and Artemis could teach hunting skills. Demeter could instruct in agriculture while the Muses would inspire people with the knowledge of the arts and sciences. Like humans, gods, too, had specialized professions. So nature and people's own future were entirely up to the goddess Fate and all the gods in general. Excluding immorality, it was in a way like a Disney Tinker Bell movie: different processes in nature (e.g., the changing seasons) are carried out by different types of fairies of specialized professions, like the tinker fairies, the winter, warm, water, garden, light, frost, plant, animal fairies, and so on.

To the contrary, the mystery religions (especially the Orphic) promoted an entirely different outlook. The worshipper was an intellectually and ethically evolving individual with greater personal responsibility for his own future including the *afterlife*. He held a deep conviction that he had certain control over his knowledge and actions; he was hopeful that, through mystic rituals and asceticism, divine immortality and wisdom were also humanly achievable and thus not an exclusive privilege of the gods. Such change of religious attitude was indicative of an unsettled, curious, open mind, one unsatisfied with the passive and strictly dogmatic mythological worldview of the Olympian religion as a means to understand nature, life, even death, and in search of something more profound and meaningful—something rational, natural, objective, universal, even humanly controllable.

In particular, during the Mysteries the worshipper ate, drank wine, danced, rejoiced, sorrowed, and felt a divinely inspired madness; he went to the extremes of frenzy hoping to experience *passion* (a physical and an intellectual suffering), *ecstasy* (etymologically, the release of the soul from the dependence of the body—recall a spirit/soul could enter a body, i.e., in-spire it, but also escape from a body, that is, undergo ecstasy), and ultimately *enthusiasm* (unification with the honoring god). With such intense emotional arousal the worshipper behaved in ways different from those of the everyday: free from the daily inhibitions and oppressions he was his true self. Simply put, during the Mysteries the believer took matters in his own hands and tried ritualistically to feel and act just as he thought his god did and hoped that life would imitate the ritual, an idea as old as religion itself but now with a new twist, an intellectual expectation by the worshipper. This ritualistic emotional enthusiasm (this potential unification with a god) led to the belief that everything divine—immortality, omnipotence, bliss, even the godly mystic (secret) knowledge—was humanly attainable, *yes, even the godly knowledge* (the secrets of nature)! The believer, of course, first had to achieve absolute purification of his fallen soul through a system of complex sacraments. Once, according to Hesiod's *Works and Days*, during the Golden Age, the first-generation people had pure souls and were allowed by gods to live with them in bliss. But people sinned and were separated from their gods. Still, until they join them again, divine aspects, such as godly knowledge, could begin to be experienced by the worshipper, so he believed, at least ceremonially during the Mysteries, via proper soul purification, prayer, the reconciliation of the gods with offerings and sacrifices, and perhaps through sympathetic magic: "If I manage to feel and act the way I think my god does, I hope I will then begin to become like one." Anthropomorphism, recall, evolved from imagining merely the *gods* in the image of man, to imagining zealously *man* in the image of the gods, a radical reversal in human psychology indeed: that is, man aspired to be like the gods: almighty, all-knowing, and immortal.

PHILOSOPHY BORN OF RELIGION

But such belief *did not* remain only ceremonial; to the contrary, it was gradually affecting even the daily life of the Greeks of Mysteries. In particular, the Orphic, who aimed for spiritual drunkenness (so wine was used only symbolically), believed that the exercise of a proper ascetic way of life could ultimately purify one's soul, elevate it to the otherworldly heights of the gods, and thus release it from its cycle of constant deaths and rebirths (thought to occur through metempsychosis). He believed, that is, that he could reach the state of apotheosis (become godlike in every aspect: power, wisdom, happiness, immortality) and thus be allowed once again the honor of eternal bliss and absolute knowledge alongside his gods—the goal of the mystery religions, which were after all religions of salvation. And so death was no longer the terrifying, hopeless, and gloomy place of Hades (etymologically, of the "Invisible," thus supposedly unknowable, where the soul is powerless and in oblivion), as in the Olympian religion, but rather a hopeful state of existence at the Elysium (the Island of the Blessed, of heroes and gods, where the soul could be immortal, conscious, free, and with divine wisdom). Interestingly, modern scientists, through their efforts to figure out the laws of nature—of creation, one might say—have in a sense similar one of the Orphic aspirations: to know the mind of the Divine.

That offered the Orphic, and generally the Greeks of Mysteries, hope for the future (including the afterlife), and in their searches for ways to satisfy their changing religious needs and fulfill their goal, they were stimulated for a philosophy of life (for the search of a deeper truth, about existence, death, nature); a philosophy born of religion. *That* was an intellectual turning point for if "philosophy . . . is something intermediate between theology and science,"[4] then the discovery of science was the expected logical next step.

This however took place only when the religiously, philosophically, and morally evolving ancient Greek, who desired the longed-for divine knowledge at any cost (through the risk of a Promethean-like retribution in the Olympian religion or through rituals and/or asceticism later in the mystery religions), grew impatient waiting for the gods to decide to (literally) *in-spire* him. And he had, at some point, realized that mystic rituals, sympathetic magic, reconciliations of anthropomorphic erratic gods in hopes for an *inspiration* (of knowledge "handed out" via godly revelation), and generally religious dogmatism or asceticism, were not working out. But what could work was simply the free and rational critique of both, nature *and the worldview of one another*—that is, learning, he thought, should come from thinking persons themselves, not from Fate or rituals, and so finally and without fear he began to imitate the actions of his favorite cultural hero Prometheus, who stole the godly secret of fire at will. And when the ancient Greek tried it he found out it was the only thing that worked. *That* gave birth to science. This "rational critique" attitude, in search of the truth, is of course useful not just as a way to do science but as a general way of life as Socrates would have attested.

Additionally, there was neither an official religion in any of the Greek city-states, nor was there a written religious corpus to which a city or an individual had to conform, nor an organized priesthood to impose particular dogmas, rituals, or a lifestyle. Consequently (and unlike the analogous case of other civilizations), Greek religion not only did not interfere with attempts for a non-mythological worldview but in subtle ways it promoted them. To understand the nature of the world they lived in humans had to think for themselves, an action, which I think, unavoidably forced them to study nature rationally. In fact, because there was no theocracy in Greece, neither political law nor morality came from the gods (or priests or kings) but from the people themselves, the type of people whose idiosyncrasies are studied in the next chapter, and who also invented democracy and pursued moral philosophy.

CONCLUSION

Human knowledge was initially thought decidable only by the gods, then hoped for through rituals and/or asceticism, but at the end proved obtainable only through one's own reason. And so in the vastness of existence, earth was only another planet, the sun just another star, neither was a god or the center of the universe. All things (including human beings) were composed of the same primary substance/s and obeyed common natural laws (or one grand law), which should be describable mathematically. Advanced life-forms evolved from simpler ones, and neither illness was caused by demons nor eclipses by gods.

The idea of intellectual progress captivated the Greeks so intensely that for the sake of knowledge they risked angering their gods, "stole" their fire and their (nature's) other secrets, and gave birth to science. But why were they able to do so?

CHAPTER 6

THE BIRTH OF SCIENCE

INTRODUCTION

What led to the intellectual transition from mythology to science 2,600 years ago, and why did this cultural phenomenon begin to unfold first in ancient Greece? The factors that are generally accepted as having created favorable conditions for such a transition were geographic, economic, religious, and political. In this chapter I add to the usual list of factors three new ones: the power of the Greek language, the effect of making a habit of scientific thinking, and the ancient Greek idiosyncrasy. The study of these factors will help us understand why science may be regarded as being born out of the Greek civilization. We begin by reviewing first the commonly accepted factors.

GEOGRAPHY, ECONOMY, RELIGION, POLITICS

1. Geographic: Locally, a landscape of natural boundaries—mountains separating cities, and the sea separating islands—helped in the formation of relatively isolated city-states (a thousand or so) and promoted intellectual diversity. Diverse ideas were ultimately shared and improved when people moved and interacted. Globally, the crossroads location of Greece exposed its people to ideas of other great civilizations from Europe, Asia, and Africa. Moreover, Greece's long coastline and many surrounding islands resulted in the establishment of coastal- and island-cities and made Greeks seafaring people. But their sea adventures aided them in demythologizing the phenomena of nature and stimulated them in conceiving rational explanations.

2. Economic: Average people became technologically inventive to better

their lives. And even though technology is not science but the application of it, technology can lead to abstract theorization about how it can be improved and consequently the discovery of laws of nature upon which technology is based. On the other end of the economic spectrum, well-to-do people used their leisure to philosophize and theorize.

3. Religious: Contrary to theocracy and hierarchy, which impose dogmatic thinking, restrict inquiry, and impede progress, religious freedom in Greece allowed for contemplation of diverse views and created a potential for betterment.

4. Political: Social freedom and democracy prompted free debates on just about everything, resulting in the conception and exchange of new and improved ideas.

Because these factors have been contemplated extensively in the literature,[1] my focus in this chapter will be on the three influences that have not been sufficiently appreciated.

The first influence is the Greek language itself. The notably communicative nature of ancient Greek helped in the conception and diffusion of knowledge in the most efficient way possible. While the first alphabet was Phoenician, the first alphabet to contain vowels was the Greek. With this innovation Greek became the first easily read and written language of the world, and the facility of written Greek became significant in the evolution of ideas and the birth of science. The second influence is simply the force of intellectual habits. Using ideas from the theory of biological evolution, I will argue that the good habit of the pre-Socratics to practice science imposed an epistemological kind of natural selection by promoting intellectually favorable environments where learning science could continue to happen and new scientists could exist, thrive, and become abundant, contributing therefore to the constant development of the scientific outlook at the expense of the mythological one. The third influence is the idiosyncrasy of the ancient Greeks: they were rational, passionate, and excessive. But it was the proper moderation between passion and logic that allowed them to become *creatively* excessive.

LANGUAGE

What first interested me in investigating the language factor was a brief statement by Nobel laureate Bertrand Russell (1872–1970): "The Greeks, borrowing from the Phoenicians, altered the alphabet to suit their language, and made the important innovation of adding vowels instead of having only consonants. There can be no doubt that the acquisition of this convenient method of writing greatly hastened the rise of Greek civilization."[2] Although the Greek language is usually not regarded as a factor that created favorable conditions for the birth of science, I will argue that its influence was subtle but profound and thus cannot be overlooked.

I will first lay the groundwork, in the next two subsections, by contemplating the general effectiveness of language in human survival and intellectual evolution. Then, in the third subsection, I will link directly the influence of the ancient Greek language on the birth of science.

The Sound of the Fittest

From the family tree of biological evolution the more anthropomorphic primates (the hominids, species that are more human than ape) are a family of species whose first member is believed to have evolved some seven million years ago.[3] Its two most recent members, who are relevant to our consideration of the effect of language on both our physical survival as well as our intellectual evolution, are the evolutionary cousins *Homo neanderthalensis* (Neanderthals) and *Homo sapiens*. Both species are thought to have evolved only about two hundred thousand years ago, with Neanderthals preceding. So at one time the two cousin species shared the earth and possibly interacted.

Neanderthals are our closest genetic relative. Physically, in some very general terms, the two species were not that different—a visit to either the American Museum of Natural History in New York City or the Smithsonian National Museum of Natural History in Washington, DC, where artists and scientists reconstructed Neanderthals' possible appearance from various findings including fossilized bones, will convince anyone of this.[4] Neanderthals were short and stocky with a more elongated skull, and *Homo sapiens* were taller and thinner with our characteristic high-dome skull. Furthermore, because the two cousin species share several brain similarities, it has been speculated that they were of comparable intelligence. This hypothesis, however, is the subject of current contention.[5]

With such general similarities, both species would have been expected to survive, but only *Homo sapiens* have managed. Unfortunately, between twenty-five thousand to thirty thousand years ago, Neanderthals became extinct. The theories for their extinction vary and are hotly debated. The cause might be just one or a combination of several, such as climate change or an isolated existence in clans, which might have resulted in limited exchange of ideas and thus a slower rate of intellectual progress than needed for surviving life's constantly changing challenges.[6]

One theory of extinction relevant to our discussion on the importance of language in survival is Neanderthal-human competition. Such competition might have been destined to be biologically unequal. For through a mutation (a purely chance change in the genome, the hereditary substance) *Homo sapiens* were accidentally gifted by nature with an anatomy comprising a more efficient larynx that could produce a richer variety of sounds, creating the potential to develop a relatively more advanced language than that of Neanderthals. This must have aided in the general survival of *Homo sapiens*. But some experts hypothesize that in a more specific way, this also might have been a contributing factor in our survival at the expense and general extinction of Neanderthals, by giving us a competitive advantage. It is probable that a better capacity for language enabled *Homo sapiens* to communicate essential survival skills such as hunting and gathering, making and refining tools, finding shelter, making friends, living together in extended social groups, forming alliances, trading, and generally learning from each other.[7]

Consequently, *Homo sapiens* developed a better understanding of the world around them and achieved an intellectual edge over their cousins the Neanderthals in all aspects of their competition. But during the early competitive environment of predators, limited resources, and in general a nature where survival was of the fittest, such intellectual advantage achieved through language skills (regardless of how primitive initially) made a difference between life and death. Thus, this theory holds, *Homo sapiens* secured their survival by overpowering and driving their own cousins to extinction.[8]

Language is a useful skill, possibly the most powerful of humankind, not only in the struggle to survive but also in our efforts to thrive and live fully. Language controls the flow of information and creates the potential for knowledge. But how rapidly does intellect evolve with the influence of language, especially an evolving language?

Biological versus Intellectual Evolution

The effectiveness of language can be appreciated further by comparing the time required for the extremely slow biological evolution of the anthropomorphic family of species with that of the immeasurably faster intellectual evolution of the only species that managed it, *Homo sapiens*, and trying to explain the reason for such a huge time difference.

Specifically, on the one hand, the biological evolution of this family describes a seven-million-year process (from its first member species, the *Sahelanthropus tchadensis*, believed to have evolved about seven million years ago, to its last and only extant member, *Homo sapiens*, who evolved about two hundred thousand years ago), but on the other hand the incredible intellectual evolution of this entire anthropomorphic family is due exclusively to the achievements of just this last member species. And depending on what might be regarded as advanced knowledge, such evolution can be condensed to an unbelievably small time interval. It could be thirty thousand years (since splendid art was painted on cave walls by Ice Age cave dwellers); or ten thousand years (since the end of the last glacial period, which roughly coincided with the transition from the lifestyle of hunter-gatherer to farmer, urbanization, and consequently the birth of civilization); or about five thousand years (since the beginning of written history when Sumerians in Mesopotamia invented the first type of writing in the world at 3100 BCE); or 2,600 years (since the birth of science); or some five hundred years (since the rebirth of science with the contributions of Renaissance astronomer Nicolaus Copernicus [1473–1543]); or three hundred years (since the Industrial Revolution); or, even more impressively, a mere few decades (since the discovery of the computer)!

To emphasize the unprecedentedly rapid cultural and intellectual evolution of the last few decades, I recall a comment by noted science fiction and popular science writer Isaac Asimov concerning the conclusion of his *Chronology of the World*: that, while his initial intention was to write the entire history of the world, from the big bang (the event that gave birth to the universe, as we will learn in subsequent chapters) to the date his book would be completed (a fifteen-billion-year period for him then), he was finally forced to conclude it with the events of 1945 instead of 1989, the book's completion date, falling short of his initial goal by a mere forty-four years.[9] And he explained that the reason was that the changes brought about by the evolving human culture between 1945 and 1989 were so many, rapid, and universal that to be effectively described would require their own book as extensive as *Chronology of the World*! He said this in 1989. Can you imagine what he might have said today, especially after the explosive evolution of the Internet? In the 1980s the Internet was just being born.

I concur with Asimov's assessment and base my understanding on the evolving notion of language itself. For from the simple sounds and symbolic cave art of the distant past to the rich languages, modern mathematical symbols, and sophisticated electronic communications of the present, language has been evolving diversified and creative new modes that allow for better conception, dissemination, and improvement of knowledge and have thus been transforming our species intellectually faster than ever before.

More precisely, with time and as a consequence of advances in mathematics, science, and technology, the notion of language has been broadened. Mathematics has added a versatile variation in symbolic and quantitative communication, while science has enhanced our imagination and invented naturalistic and rational interpretations of nature, and technology, mainly after the invention of computers (especially their interconnection via the sociologically innovative Internet), has enriched communication through myriad modes, including ones that affect all people of this planet and potentially intelligent beings of other star systems. Traveling at the speed of light, a radio signal transmitted from the Arecibo Observatory in Puerto Rico in 1974 has as its destination the globular cluster M13, a group of some three hundred thousand stars in the constellation of Hercules twenty-five thousand light-years away from us.[10] The signal's coded information about us can be easily decoded if intercepted by an intelligent alien life-form.

For millennia, the idea of language has included more than gestures and sounds. Knowledge can be recorded many different ways and in places other than the human brain. Thus while we no longer need to remember everything, everything can still be remembered because the knowledge of the past is readily available and therefore accelerates the rate of progress. One can learn the accumulated knowledge of millennia by simply reading a book!

And all of this can take place because we are anatomically able to speak sounds, are instinctively curious to develop them into coherent language, and are intellectually successful in habitually passing on such great skill to our offspring. And such is the power of language: it is a skill for rapid and extraordinary intellectual bursts! Unquestionably language has been aiding in the advancement of science. But did it aid in its birth?

Ancient Greek Language and the Birth of Science

The evolution of the Greek language has been a huge topic for scholarly research. While I admit ignorance on such an immense linguistic field, I also know the generally accepted facts about Greek's extraordinary richness, such as a plentiful vocabulary, thorough and rigorous grammar, diverse phonology, and successful orthography (i.e., spelling), all of which contribute to the language's highly expressive and communicative nature. This distinct nature leads me to contemplate the connection of the language and the birth of science. But first some history.

Spoken since at least 2000 BCE and written since at least 1400 BCE (not yet with the Greek alphabet, which evolved a few centuries later), Greek is one of the world's oldest recorded living languages and the longest documented from the Indo-European family of languages where it belongs. Phoenicians invented their alphabet around 1050 BCE. Modeled after that, the first true alphabet containing vowels was invented by the Greeks around the eighth century BCE. It was rapidly diffused throughout ancient Greece. With this innovation Greek became the first most easily read and written language of the world. This is so because alphabets are phonetic: each different sound of a language can be represented with a unique symbol, and thereafter symbols can be combined to write and sound all the words of the language. Therefore, with an alphabet every language can be written and read relatively easily. In contrast, a pictographic writing system, in which a picture represents a word or phrase, is more complex. The success of the Greek alphabet is also indicated by the fact that after some three thousand years, Greek is still written with the same letters that served as a basis for the Latin letters, and which, in turn, have been the basis of several modern languages. While Greek has been evolving, its overall identity has been basically preserved. Greek has remained relatively the same language until today, a rather rare but not accidental linguistic phenomenon. Parenthetically, part of the explanation for this might be that the Greeks value so highly the written works of their ancestors that their references to them kept the essence of their language unchanged.

Because of its simplicity, the Greek alphabet assisted in making the good habit of literacy accessible to all in ancient Greece. By the fifth century BCE every male citizen was expected to know how to read and write.[11] "The Greeks founded such an eminently literary culture."[12] Such widespread literacy undoubtedly accelerated progress. In contrast, the complexity of some other cultures' writing systems, often combined with their theocratic (and hierarchical) political systems, made writing the nearly exclusive privilege of priests and professional scribes and not the populace, a situation arguably unfavorable for developing science. Greek literature begins with Homer's monumental epic poems the *Iliad* and the *Odyssey*, dated by consensus from around eighth century BCE. However, their surviving present form is at latest from sixth century BCE, the century when Greek philosophy, science, and mathematics began. From around 700 BCE are Hesiod's poems *Works and Days* and *Theogony*. All four works were significant in educating Greek youth.

These chronological facts indicate that Greek's relatively early growing richness was present by the time of the birth of science in early sixth century BCE. This evidence, together with the fact that Homer was from Ionia—which was also the birthplace of the first scientists (the natural philosophers Thales, Anaximander, Anaximenes, Xenophanes, Heraclitus, Pythagoras, and Anaxagoras)—proves that science was born at a place and time where language was already advanced enough to aid the evolving scientists in the clear articulation of their theories.

This is a significant conclusion, for it links directly the positive influence that the ancient Greek language had on the birth of science. Greek had equipped the early philosophers with the skills for conceiving and formulating their abstract thoughts, clearly expressing their minds, and efficiently converting their raw intelligence to systematic, rational, transferable, and debatable knowledge. Without such a productively expressive language, their scientific theories would have remained unrefined, perhaps not even conceived in the first place.

A poor language reduces not only the ability to express oneself but also the potential to learn from others, for if neither we nor others can think and communicate clearly, we can neither influence nor be influenced. And the poorer the speech and writing acquisition are, the more inadequate the cognitive process becomes.

It seems no accident that the Greek language had been maturing roughly simultaneously with Greek thought in philosophy, science, and mathematics. The sounds and symbols of a communicative language could create clearer thoughts, which could then refine further the language in a continuous interactive cycle of the evolution of both. But mathematics is also a form of language, most particularly the language of science. So while by language we usually mean the communication in terms of sounds and written words, mathematics has tremendously empowered such new ideas by utilizing numbers, equations, complex diagrams, and abstract concepts. Mathematics helped to develop abstract thinking and to quantify science. In turn, science enhanced technology, which in turn enhanced both science and mathematics, in a mutually productive process. Now since mathematics adds a valuable extension to the definition of language, can we find yet another link between language (specifically the mathematical) and the birth of science?

During the rise of Greek civilization, science and mathematics were driving each other and evolving simultaneously. The first natural philosophers were both scientists and mathematicians. Russell has said, "The preeminence of the Greeks appears more clearly in mathematics and astronomy than in anything else."[13] Mathematics was a skill that enabled them to conceptualize and more easily make their scientific theories rational; but equally important, their unprecedented physical intuition concerning the workings of nature aided them in advancing mathematics and thus the language of science.

Thales (who flourished in early sixth century BCE) was also a geometer. After him, the Pythagoreans were superb mathematicians and the first to implement the mathematical analysis of nature, a practice of vital significance in modern theoretical physics. Physicist and Nobel laureate Erwin Schrödinger (1887–1961) argues that what guided Democritus (the last of the pre-Socratics) in conceiving his atomic theory of matter was his deep insight of mathematics.[14] In fact, the most enduring discoveries from Greek science of antiquity were by natural philosophers who were also accomplished mathematicians.

The mathematical knowledge that was a common characteristic among most of the pre-Socratics seems to indicate that science could not have been born by persons who did not know the language of mathematics. This is yet another conclusion that links directly the positive influence of the ancient Greek language, which in its broader definition includes mathematics, with the birth of science. Without a doubt, the clear conception and coherent expression of complex ideas were made easier by the communicative nature of the prolific ancient Greek language. But could the scientific birth have survived and matured without good habits?

HABITS

A combination of factors aided in the emergence of the first natural philosophers and in the transition from mythology to science. This unfolding new knowledge gradually advanced, and spread, grew popular, respectable, and of practical value but also abstractly meaningful and satisfying. Among the Greeks generally, seeking knowledge became a way of life, a scientific habit that characterized the culture. And even though acquired properties such as knowledge and skills are not biologically inherited, habits (such as practicing science) and behaviors (such as a desire to advance the scientific outlook) associated with such properties are transmittable culturally through teaching and can still change the environment in complex and subtle ways. And in turn, through the process of natural selection from biological evolution, the environment can influence a species by controlling the direction of its evolution.

Specifically, the good habit of the first natural philosophers to practice science imposed an epistemological kind of natural selection by promoting scientifically favorable environments where learning could take place and new scientists could exist, thrive, and contribute to the constant development of the scientific outlook at the expense of the mythological one.

But since my goal is to explain the critical role that habits play in our intellectual evolution from the point of view of biological evolution, I first need to discuss further the notion of natural selection imposed by a habit.

Imposed Natural Selection

The process of biological evolution of the species begins with a mutation (a random alteration in the genome that can result in a new hereditary characteristic) and continues with the mechanism of natural selection (which says that inheritable characteristics that are also environmentally favorable become more common in successive generations; hence it describes the role of nature in the preservation or extinction of a species). Natural selection can proceed as a consequence of a variety of environmental influences such as chronic periods of coldness, hotness, dryness, wetness; the eruption of a super volcano; changes in atmospheric composition; an asteroid-earth collision; radiation from the sun; or a supernova explosion.

But natural selection can also be imposed by the habits of a species; after all, species are part of nature and their actions affect it. In this case, if some members of a species already have or develop an inheritable trait (a mutation) that is favorable to a kind of environment created by a habit—either their own or another species'—then they will be naturally selected. This means that these members will begin growing up more easily, prospering, adapting, preferentially reproducing, and becoming more abundant in such an environment that is friendly to their rare trait. Assuming the habit persists, in time the species will gradually evolve to the point that most of its members possess the genetic trait favorable to the environment created by the habit. Let's look at two specific examples:

1. Microbes: While on the one hand a moderate use of antibiotics can be beneficial to our survival by killing myriad common but still harmful microbes, on the other hand a habit of thoughtless overuse of antibiotics can promote the evolution of rare but much more dangerous microbes (superbugs) that are resistant to the antibiotics we use. Natural selection, in this case imposed by the habit of overuse of antibiotics, can make common population characteristics

rare (common microbes get killed) and rare ones common (mutant microbes resistant to our antibiotics get multiplied).

In the microbes example, the habits of one species, humans, can impose natural selection onto another species, microbes. Humans actually have since urbanization been imposing natural selection onto several species; the plants we have been domesticating, the animals, even species with which for one or another reason we must systematically interact—like the aforementioned microbes.

Equally fascinating is another fact of evolution: that the habits of a species can also impose natural selection onto itself—for example, humans can impose natural selection onto themselves! Before I discuss humans, let me first discuss briefly the birds.

2. Birds: With the desirable genetic trait of wings, birds avoided many predators only when they began habitually using their wings for flying and building their nests high up in trees. Such a habit imposed natural selection by creating an environment that selected and promoted even further the evolution of birds that could fly the best. With time these skilled high fliers became more abundant, while birds that could not fly proficiently became rarer.

The mythological worldview was once popular and the scientific rare, but since the birth of science their status has been gradually reversing, a fact that is contributing to the overall intellectual evolution of the human species. This observation brings me to a hypothesis, to be introduced in detail in the subsection below, that the good habit of doing science imposes an epistemological kind of natural selection that gradually selects people with scientific and, in general, intellectual tendencies. Such a habit not only secured the safe birth of science during the critical early stages, 2,600 years ago, but also has since then been contributing to the overall evolution of the scientific outlook at the expense of the mythological.

Erwin Schrödinger in his *What Is Life? & Mind and Matter* gives a detailed analysis of (a) how behavior in general influences natural selection and thus the process of biological evolution and (b) how our invaluable characteristic of intelligence allows us to conceive and implement incalculable choices and so both our behavior and consequently our evolution depend on us, at least to a certain degree.[15] Thus, he argues, our evolution does not depend solely on chance mutations. This is an encouraging but also challenging prospect. Based on these two points, he speculates on the potential of intellectual degeneration in our species. Below I will focus on an analysis exploring the opposite: how practicing science habitually has imposed an epistemological natural selection and has been influencing positively the evolution of the human intellect. (This is not to say, of course, that it could not influence it negatively.)

Habits Influence Evolution

Since habits can impose natural selection and cause biological evolution, they can also cause intellectual evolution, for our organ of intelligence, the human brain, is just one of many body organs known in biology to have been evolving. So good human habits can cause a biological evolution of the brain and consequently create the potential for intellectual evolution. Just as birds that could fly the best were selected in the environment where flying became a bird habit, it is not unreasonable to suppose that the developing good habits of the pre-Socratics to understand nature scientifically instead of mythically imposed natural selection by creating favorable environments for new scientists to flourish, multiply, and evolve. In short, the good habit of doing science set up an epistemological environment where the scientific man, in general the intellectual man, or, to say the least, the man of scientific appreciation, is favored and thus naturally selected.

The pre-Socratic era was the first habit-forming period for science. Specifically, several good habits of pre-Socratics—people who were keenly observant; curious; skeptical; investigative; unconventional; open-minded; free-spirited; innovative; rational; passionate; critical; philosophical; eager to speak, write, and debate; truly scientific; and generally epistemological (interested in knowledge of diverse fields)—have been inherited by succeeding generations, from their place to another, from the few to the many, from then to now, from ancient Greece to the rest of the world and seem to have been imposing an epistemological kind of natural selection by promoting scientifically favorable environments.

These good habits have therefore contributed systematically to the formation of an ever-improving scientific worldview at the expense of the mythological one and consequently advancing our overall intellectual evolution. For truly epistemological individuals have found such environments intellectually appealing, rewarding, welcoming, and increasingly more adaptive, so much so that today's humans have evolved to become intellectually superior to our ancestors, in fact to any other species known. Hence the kind of environment set up by the good habit of learning (or flying, in the bird example) favors, through imposed natural selection, the increase of those interested in learning (or flying).

A Good Genetic Trait and a Good Habit

So, having an environmentally desirable genetic trait (for example, a larynx, a complex brain, legs, wings) from which a good habit can develop (language, learning, walking, flying) is only one required element in the struggle for survival. Using the trait systematically and habitually is the second required element. For only then can the trait influence the environment via imposed natural selection so that the members who have it can be naturally selected even further and consequently increase their chances for survival and betterment by becoming environmentally more fit. In the evolution of birds, for example, those that did not take up the good habit of flying, despite their anatomic ability to do so, generally have less chance to survive attacks by predators. On the other hand, the expert fliers that use their wings proficiently tend to flourish. Varieties that are not environmentally favorable (such as birds that despite having wings are not using them) can become rare and perhaps extinct. But even if they do manage, not following good habits makes their existence much more vulnerable.

Extending the logic of the bird example into the realm of humans and their intellectual habits, it can be argued that a greater chance to flourish belongs to those who use their brains intelligently and try to develop good learning habits, such as attending school, in order to keep up with new challenges and opportunities of a fast-changing environment. In this bird-human analogy there is, however, an important difference in favor of humans. We have a far greater level of intelligence. We have a choice of how to behave, and since behavior influences evolution, through our choices we, too, contribute significantly to our own evolution. Specifically, through chance mutations we were endowed by nature with the raw intelligence of an anatomically complex brain, but what also plays a critical role in its development is our conscious choice of using it productively. Again, I assert that practicing science habitually has imposed an epistemological kind of natural selection, has changed the intellectual environment, and has allowed us to realize our potential to live up to our name and become truly *sapiens*: wise. Starting around the sixth century BCE, science, philosophy, and mathematics were gradually becoming a way of life in ancient Greece, increasingly systematic and habitual, not just for a few individuals in a few places but for whole populations in many cities, especially in the education of the young, creating therefore a better chance for this way of life to be passed on to future generations and to people in new places. Since then, because learning science has gradually become a significant skill in life, the numbers of those with the mythological worldview have been decreasing, while those who are scientific (rational) have been multiplying. This development is comparable to the declining numbers of the rarer birds that cannot fly proficiently and the growing numbers of the numerous expert high fliers that flourish in an environment where flying became an important skill for survival.

Practicing Science Habitually

The notion of a habit was crucial in the development of Greek civilization. There is absolutely no reason to believe that, before or after the Greeks and independently of them, others in the world would not have conceived a scientific idea about nature or a mathematical demonstration of some theorem. In fact, the proof of my opinion is that all kinds of people from all over the world do, or can learn to do, science and mathematics. But if something profound like practicing science had not happened habitually, such good skill would not have spread and perhaps soon would have vanished without significant effect on society. But evolutions occur when a phenomenon leaves a mark on the environment. A significant reason for the rise of Greek civilization was that philosophy, science, mathematics, and the love of free thinking—and consequently democracy, which aided in the preservation and continuation of the good habit of practicing science—all evolved into a good habit that has been influencing the world ever since.

Hence, it must be acknowledged that in addition to a variety of more commonly understood factors, an important element in the intellectual transition from mythology to science in ancient Greece was that the Greeks, starting with the pre-Socratics, pursued their new ideas in a systematic, persistent, and habitual manner. And in their explanations of how nature works, the pre-Socratics applied exclusively the scientific outlook for all phenomena of nature. For them every phenomenon had a natural cause; thus supernatural interventions were ruled out. This is the reason why Greece, from roughly the sixth century BCE, may be considered the birthplace of science. Had the pre-Socratics explained some phenomena naturalistically but others supernaturally, this birth of science would not have occurred. From the Greeks the scientific outlook spread and today is a way of life and a culture, a human culture.

IDIOSYNCRASY

Was it geography that broadened the Greeks' intellectual horizons, or was it their curious, adventurous *open* mind that led them to see the world (nature, their own culture, and that of others, as well as human itself) with a critical eye—freely questioning and analyzing it for, like Socrates, they thought "the unexamined life is not worth living" and so they sought for more objective, coherent, universal, and timeless truths about nature (including humans)? Was it tools (or free time) that sharpened their intellect, or their intellect that sharpened the tools and pursued leisure time to philosophize? Was it the absence of theocracy (which imposes dogmatic and mythological thinking) that promoted free and rational thought among the Greeks, or was it their tendency for free and rational thought that opposed the formation of theocracy? Was it democracy that encouraged free public debates and open dialogues, or the Greeks' critical mind and love for freedom of speech and independent thinking that contributed to the invention of democracy? Was it the advanced Greek language that aided their thoughts (to be clearly conceived, expressed, and disseminated), or their thoughts (formed out of love for self-expression, rhetoric, dialectic, literature, and inquiry—for *any* subject was to them open for contemplation, debate, and criticism) that aided the advance of their language? Could causes be mistaken as the effects? Although not absolutely, to some degree, yes, they could. If so, then in the rise of Greek civilization the idiosyncrasy of the Greeks cannot be overlooked. Well, then, what kind was it?

Rational, Passionate, Excessive

Plato's dramatic chariot allegory (from his dialogue *Phaedrus*) captures the essence of the ancient Greek idiosyncrasy.[16] According to the allegory, a charioteer, who represents (personifies) the human soul, tries to drive a two-winged-horse chariot to the proper destination, that of truth. But the journey is not easy because the horses, which represent the complex twofold nature of the soul, have excessive energies but competing tendencies and pull forcefully toward their separate ways. One horse represents the tendency of the soul to conform to orderly reason, the other to surrender to defiant passion. Reason can tame passion, but passion can cloud reason. Reason alone may lead the soul down a safe but dull path, while passion alone may lead it down a risky but uncommon one. Nonetheless, it is not the one or the other tendency that is the good or the bad for the soul but their combination. Destination Truth (the discovery of something extraordinary) can be reached only by harnessing properly both horses' excessive energy. For not only reason is natural and needed but so is passion. The ancient Greek idiosyncrasy, like the charioteer, was a synthesis of the rational and the passionate nature. And each nature was excessive and at war with the other, like the mighty horses that pull vehemently against each other. But their delicate harmonization, whenever it was managed, is what steered the Greeks to creativity.

To eliminate the backward passions of magic and superstition and to liberate humankind from its dependence on irrational supernatural forces, the ancient Greeks rationalized nature and gave birth to science and intellectual freedom. To fight the corrupt passions of tyranny and autocracy they figured out the law and gave birth to democracy and political freedom. To face the blinding natural passions of human soul they contemplated morality and criticized themselves, each other (in fact, welcomed the latter), and the policies of their cities—and all in public—while giving birth to moral philosophy and philosophical freedom. It took a rational nature for them to philosophize, politicize, and do science; as well as to, or actually try to, "know thyself" (to have self-knowledge, to be conscious of their limitations and potentials) in order to take responsibility and control over their own understanding, actions, and future (away from tyrannies, hierarchies, superstitions, dogmas, and even their own passions). But also it took a passionate nature for them to aspire and choose freely in spite of consequent suffering, even death (as the heroes in their tragedies); to choose to die for freedom by battling internal oppressing tyrants and external despotic kings (as the heroes of their wars); and to despise powerful capricious gods and age-old mythical traditions and mysticism for the sake of religion based on philosophy and of knowledge based on rational inquiry (as the heroes of their everyday life). But none of these pursuits, neither religion nor philosophy nor science nor democracy nor freedom, were achieved through reason or passion alone. It was not just the one or the other that was the good but their combination. The Greeks were passionately rational but also rationally passionate.

Russell has written of the Greeks, "Without the Bacchic element [passion, ecstasy, enthusiasm, frenzy, impulse, spontaneity, suffering, sorrow, joy, a divinely inspired madness, the liberation from the constraints, agonies, and pressures of the everyday, and the expression of one's true self, all feelings aroused at the Bacchic, the Dionysian Mysteries], life would be uninteresting; with it, it is dangerous. Prudence versus passion is a conflict that runs through history. It is not a conflict in which we ought to side wholly with either party."[17] Plato (in his dialogue *Phaedrus*) believed something similar: "He who, having no touch of the Muses' [divine] madness [passion] in his soul, comes to the door and thinks that he will get into the temple [become a prophet] by the help of art [by reason alone, by mere knowledge of the art of prophecy]—he, I say, and his poetry [sane prophecy] are not admitted [for they lack a touch of a divinely inspired madness, passion]."[18] Russell continues, "A large proportion of them [Greeks], were passionate, unhappy, at war with themselves, driven along one road by the intellect and along another by the passions. . . . They had a maxim 'nothing too much,' but they were in fact excessive in everything—in pure thought, in poetry, in religion, and in sin. It was the combination of passion and intellect that made them great, while they were great. Neither alone would have transformed the world for all future time as they have transformed it."[19] By the way, passion was one of the "evils" sealed in Pandora's box, but Pandora's curiosity—*Greeks'* curiosity—freed it, for reason without passion is dull. At the end, apathy (etymologically, not suffering) was the evil not passion (suffering, physically and intellectually).

Their life's philosophy was everything in moderation, self-control. But it was inspired because they were excessive and not of moderation for "noble self-restraint [reason] must have something [excessiveness, passion] to restrain."[20] One wants to make self-control his philosophy because one is excessive in his psychology. Hence philosophy's goal was to warn them of the potential dangers of excessiveness, and their hope was to keep their excessive nature under control through self-control. The result, the proper moderation between passion and logic, allowed them to become *creatively* excessive, and in their search for a new worldview they invented democracy, science, and philosophy. Passion drove them away from the ordinary, but logic controlled their passion and allowed them to embrace the extraordinary. It is their zealous and continuous search for such moderation that shaped their adventurous path in history, leading to Western civilization and those non-Western that aspire to Greek ideals—great liberties such as the personal, the civic, the political, the religious, the scientific, and in general the intellectual. Incidentally, had we not had these great liberties (those of us lucky enough to have them), would we be willing to sacrifice ourselves in order to get them?

The ancient Greeks did. Nonetheless their critics say that the Greeks did not live up to their own ideals. Bruce Thornton responds that this failure, whenever it happened, "reflects only the banal truth that humans [*everywhere*] rarely live up to their own aspirations"[21]—Greeks, in other words, were humans first. And this is a consequence of the fact that the war between reason and passion—the challenge to steer successfully the "chariot," that is, our self, to the Good—is a universal human condition, a great truth that points toward yet another great truth of Greek origin—particularly a Stoic philosophy—that of our common humanity transcending our uniqueness in individuality: we are all rational beings with passions by nature, all other differences are superficial and/or accidental. Commonality, and in fact oneness, as a general law of nature (of which man, too, is part), appears (as we will see) also in Greek natural philosophy as early as Thales—all things for him are made from the same stuff, water—and attains an astounding uniqueness in the worldview of Parmenides—for only what is, is, only the Being exists for him, one and unchangeable.

Stephen Bertman asked why the Greeks were the ones who invented science. "Because," he argued, "the seminal principles of Greek civilization—humanism, rationalism, curiosity, individualism, the pursuit of excellence, and the love of freedom—were uniquely compatible with science's own essence."[22]

CONCLUSION

Geography, economy, religion, politics, language, and the practice of good habits, all appear to have an interwoven and critical role in the creation of favorable conditions for the rise of Greek civilization and the birth of science. However, it is likely that even these do not tell the whole story. For example, I believe that to understand the rise of a civilization we must also attempt to understand the idiosyncrasy of the people who caused such a rise. With this in mind, the ancient Greeks were passionate, rational, excessive, original, critical, political, religious, philosophical, scientific, and brilliant. They debated freely, zealously and with no fear, any theory to the end despite its implications. They took chances, made mistakes but rose victorious. And one of those rises was the birth of science. Their scientific theories, although the very first, were extraordinary! What were those theories and how do they measure up with our sophisticated mind-bending modern science after two and a half millennia of scientific progress? The answer will be surprising.

PART II

THE PRE-SOCRATICS IN
LIGHT OF MODERN PHYSICS

CHAPTER 7

CLOSE ENCOUNTER OF THE TENTH KIND

It is the beautiful season of summer. I have been reading this great story for hours, since dusk and during the absolutely moonless night. It is now almost dawn. I am very tired but do not want to put the book down. It is unusually original and wonderfully profound. The pages are one by one turning. I am really exhausted and am fighting sleep because the story is so good.

All of a sudden I hear a splash. The day is hot and bright, the sky blue, the sun yellowish-white. I look up and see Thales falling into the fresh, cool, flowing waters of a river. "Not very wise," laughingly remarks the atomic Democritus, staring curiously in the void with undivided attention. But Thales is carefree and zestful, having the matter of water primarily under control, getting up but purposely jumping back in again, with novel, childlike, passionate, and playful curiosity. Off from the center of the action, revolving around a Central Fire, carefully preparing his food while singing in a harmonious but almost secretive whisper, is the legendary Pythagoras. Inspired by the moment, he stops the song and begins counting the proportionally spaced ripples of the water from each splash Thales makes. He is a prominent mathematician, able to face squarely all mathematical oddities, but he is also a cosmopolitan musician who enjoys masterlike attention from orderly and exclusive gatherings of crowds. "Cosmic justice, conserve and save the phenomena," thirstily shouts the infinitely abstract but also practical Anaximander, the genius of antitheses, feeling the heat of dry air and expecting that it will soon be neutralized by the opposite coolness of wet water.

It is really a beautiful and hot day, but Anaximenes finds a creative and concrete way to moderate the heat and thereby cool down. With his lips nearly closed, he blows air out onto his body, noting that it emerges colder than when his mouth was wide open, causing his condensed sweat to rarefy and evaporate. Sitting at a distance, away from the many, boldly being where no one has been before, is the enigmatic Heraclitus, who skeptically observes the process of the constantly changing events, going through conspicuous but also subtle changes. He is quite certain he has previously taken a bath in this river's fresh waters, but then again, strangely, everything looks new and changed. What is the Logos (cause) of all these changes? To the contrary, judging all of the sense-perceived reality to be deceptive, there is the one and only Parmenides the ontologist, proud and relieved. For journeying during the darkness of night and into the light of day, through the unknown, from afar, he found the true way here by intentionally avoiding the known and opinionated way of all others.

Anaxagoras's *nous* (or intellect) finds everything to be a puzzle: "How is it that all these people from different eras of time and different places are here?" He wonders by skillfully placing his hands over his head. "Indeed a paradox, a paradox of space and time," adds the argumentative and prolific Zeno, who, through dialectic (the method of reductio ad absurdum, or reducing to the absurd), is trying to prove that motion is an illusion of the senses, and so no one really moves, despite that all appears to so do. "Are you sure space and time are the only elements in the puzzle?" melancholy Empedocles challenges, while, in the name of episteme (knowledge) and his love for strife, holding tight onto his clepsydra (a water clock), he risks a dangerous experimental leap through the air and over the flames of fire but lands safely on earth, in fact in the water, just beside me.

"And who might you be, young fellow—the modern physicist?" he asks. As I respectfully nod in awe, I feel all eyes curiously staring at me as if I'd been expected. And immediately the brightness of the day surprisingly turns into a mysterious twilight. "I have been predicting an eclipse at your arrival," Thales says, while nostalgically shaking off primarily the substance of water from his wet, muddy, ripped, and unfashionable clothes. "We have been longing to know your story," he adds. Moments later, the bright daylight is pleasantly restored. It is now noon. "It is a beautiful day indeed," I say humbly, "for I am learning yours. And I will tell you mine, too, but under *your* sunlight, for eclipses are ephemeral and pass, but your knowledge is timeless. You *still* bring fire to modern science." The day is still young, and who knows of the morrow?

Everyone's senses are keen, observing the changing sights, listening to curious sounds, smelling soul-awakening aromas, tasting the sweet air, touching the cosmic elements. But so is everyone's intellect contemplating it all. What a beautiful day! What a beautiful nature! What is her nature?[1]

CHAPTER 8

THALES AND SAMENESS

INTRODUCTION

Thales (ca. 624–ca. 545 BCE) was interested in how nature works. He was the first to ask what things are made of and what the properties of matter are. These are still the most fundamental and difficult questions of science. His answers were based on solely rational arguments, uncluttered by myths, superstition, rituals, or the actions of capricious gods. His approach was therefore the same as that of modern science.

He reasoned that in spite of the apparent diversity and complexity in nature, all things are made from the *same* stuff: water, and all things obey a common set of unchanging basic principles, water's transformations (e.g., its solidification, liquefaction, and evaporation). Thus for Thales nature is characterized by a certain sameness or unity between all things, however diverse they may be, an overall intrinsic simplicity.[1]

THE EARLIEST SCIENCE VERSUS THE LATEST SCIENCE

While the primary substance of matter is not water, what is the primary substance had not yet been discovered. Nonetheless, presently, according to the standard model of physics (introduced further in the section titled "Sameness"), the building blocks of matter are microscopic particles called quarks—constituents of protons and neutrons—and leptons—particles including electrons. And the plethora of diverse things is partly due to their transformations (from one type of particle into another), not to the transformations of water. Despite these new discoveries, still, Thales's notions about sameness—that all things are made from one and the same substance—and about the transformations of matter, are of timeless scientific appeal. His idea about the transformations of matter, in particular, not only describes a fundamental property of the modern concept of energy (or matter, since, as Einstein's special relativity theory makes clear, they are equivalent and transmutable into each other), namely its ability to transform into various forms and cause change, but also employs causality, because for him the cause of all other things is the transformation of just one primary substance. But why was water the underlying principle/cause of such sameness and unity?

WHY WATER?

Several observations might have stimulated Thales in his speculation that all things are transient forms of water. Some ancient accounts such as Aristotle's[2] and Aëtius's[3] give us some insight. Water is required for the survival and development of all kinds of life. Primitive life exists in moist environments, and animal sperm is liquid. Also, since water transforms easily into the three forms of matter, the solid (as ice), the liquid, and the gaseous (as water vapor), and into a variety of shapes, it could, Thales might have thought, also transform into everything else, such as rocks or metals. Now, while all substances transform into the three states of matter (e.g., given enough heat, a solid piece of metal can melt and evaporate), water is the only substance of daily experience that does this before our eyes and on a regular basis through the changing seasons, something that observant Thales could not have missed. Furthermore, it transforms more easily: its evaporation temperature of about 100 degrees Celsius (at sea level) is smaller than that of, say, copper, bronze, or iron—materials that in antiquity were heated and melted to make tools—so one does not need a lot of heat to vaporize it; and in the cold winter, water is the only substance to turn to snowflakes and solid ice of all sorts of shapes. So its choice as a primary substance over other things appears logical. Thales might have reinforced his water hypothesis, I speculate, from another everyday observation, namely, that when heated or burned all things release (or so it seems, anyway) water vapor. One example that might have been an inspiring clue for his water doctrine might be observing the rising smoke from a burning piece of wood mixing with air and clouds, which in turn can be mixed with rain or snow and blend with the soil on earth. At first glance, smoke, air, and clouds, are like (or seem to be) water vapor; rain and snow are water, soil contains the water and snow of the rain or snowstorm, so it might actually be thought of as transformed water, and so might then be the plants (thus wood), since, starting as seeds, plants grow from the soil and "are nourished and bear fruit from moisture,"[4] and so might also be the animals since they eat plants or each other.

Since processes of this sort appear causal with water as the first cause, then it seemed logical to assume that everything is made from the same stuff—reconstructed from the same first principle—and that, in general, everything in nature is characterized by a certain subtle sameness.

SAMENESS

Sameness is a core concept in modern physics, not only because it emphasizes a universal, underlying, simple principle as a characteristic of all things in nature, but also because it points to a commonality in their ultimate origin. Unity (in the sense that everything can be derived from one and the same principle), Thales reasoned, is a subtle, intrinsic property of nature.[5] This idea inspired all other pre-Socratics (each creating his own special theory on unity, as we will see in subsequent chapters), and, in turn, they have inspired scientists of recent times.

James Clerk Maxwell (1831–1879) unified successfully the electric and magnetic forces by proving mathematically that they are really two manifestations of the same force, the electromagnetic. The electric force is caused by the electric charge: the positive and negative. Objects of opposite electric charge attract one another while objects of the same type charge repel. The magnetic force is caused by an electric charge in motion. A permanent magnet has two poles, the north and the south. Opposite magnetic poles attract, and similar magnetic poles repel.

Albert Einstein (1879–1955) from 1925 until 1955 attempted unsuccessfully to unify the electromagnetic force with gravity—which makes things with mass attract each other, such as the earth and the sun. Gravity, still the most puzzling of the forces, has not yet been unified with any of the other three known forces of nature (the aforesaid electromagnetic, and, coming up below, the nuclear weak and strong), despite the fact that through the work of Isaac Newton it was the first force to be described by a mathematical theory: Newton's law of universal gravitation.

Nonetheless, success struck another physics front with the combined efforts of Sheldon Glashow (1932–), Steven Weinberg (1933–), and Abdus Salam (1926–1996). In the 1960s the three physicists managed the unification of the electromagnetic force with the nuclear weak force in what is known as the electroweak force.[6] The nuclear weak force is responsible for the radioactive decay of unstable nuclei such as that of uranium, and the transformation from one type of material particle into another—Thales's notion on the transformation of matter is a significant process in modern science. The experimentally confirmed unification of the electromagnetic force with the weak force occurs at high energies and temperatures—where the two forces have the same strength and are indistinguishable, thus they are considered as one force. Whereas at lower energies/temperatures (generally those of everyday experiences) these two forces are two expressions of the same force: the electroweak.

The standard model of physics is the theory that combines the knowledge of the electroweak force and the nuclear strong force—which binds the quarks in the protons and neutrons and also the protons and neutrons in the nucleus of an atom. It is the best model so far because it combines successfully several theories to explain how particles interact and how the universe works. Quarks and leptons are among several experimentally confirmed predictions of the standard model. In fact, even more importantly with respect to Thales's view, according to the standard model, the materialness of quarks and leptons—in particular the source of their mass—is a single type of particle called a Higgs boson. On July 4, 2012, scientists working at the Large Hadron Collider, the most powerful atom smasher in the world, announced that they had discovered a new particle consistent with the predicted properties of the Higgs (the Higgs is a topic to be revisited in chapter 9, "Anaximander and the Infinite" and chapter 17, "Democritus and Atoms").

Through a Grand Unified Theory (GUT) physicists hope to extend the standard model by creating an experimentally verifiable theory in which the electroweak force and the nuclear strong force are unified. Several good candidates for a GUT do exist, making concrete testable predictions (such as the decay of a proton, not yet observed), though none has so far been experimentally verified.

Finally, a community of ambitious physicists is currently on the quest for the ultimate principle of sameness, that is, for the absolute unification of all four aforementioned fundamental forces of nature in what is termed the Theory of Everything (TOE). A TOE hopes to establish that everything in nature is explainable by a single overarching principle and its associated equation, that everything is truly a consequence of just one primary substance, as Thales initially claimed. A possible TOE that is still in its formative stages and thus highly speculative goes by the name of string theory. It seeks to describe nature in terms of vibrating strings of energy in ten or more dimensions. According to string theory and in agreement with the essence of Thales's idea, everything is made from the *same* stuff: absolutely identical strings! These strings, like violin strings, have different modes of vibrations that are speculated to manifest as different types of particles, which include the quarks and leptons.[7] In string theory, therefore, it is these exotic strings that are theorized to be the primary substance of the universe.

Now, we are aware of the three dimensions of space (think of them as three edges of a cube meeting at a vertex) and the one of time—thus with respect to what we can readily experience, we live in a four-dimensional universe. The other six spatial dimensions predicted by string theory are hypothesized to be curled up into unimaginably small ball-like geometrical shapes of 10^{-35} meters (as big as the vibrating strings themselves), and thus not easily detectable. When string theory considers an eleventh dimension it becomes a more complete theory (with fewer mathematical inconsistencies) that goes by the name of M-theory. Because in M-theory two-dimensional membranes vibrate in an eleven-dimensional universe (instead of one-dimensional strings vibrating in a ten-dimensional universe), M may stand for "membrane," though it has never been specified. The extra dimensions were required to make the equations of string theory consistent. The objects that resulted from the solution of those equations were not our familiar point-like particles (e.g., the electrons or the quarks) but rather were one-dimensional strings or two-dimensional membranes. Through a TOE we hope to reduce all of nature into one fundamental substance with its transformations, an utter simplicity and sameness of Thalesian grandness.

The paramount challenge in finding a TOE is rooted in our inability so far to combine the rules of quantum theory with the rules of Einstein's theory of general relativity. On the one hand, quantum theory describes successfully the behavior of the microscopic world of subatomic particles (such as electrons, protons, neutrons, quarks, etc.), where laws of quantum probability—the central concept of the theory—replace the deterministic laws of classical physics—which includes Einstein's, Newton's, and Maxwell's theories. On the other hand, the theory of general relativity describes successfully the behavior of the macroscopic world of planets, stars, galaxies, and generally the large-scale universe by explaining how space, time, matter, and energy are all inextricably intertwined and how gravity works. (Aspects of both of these theories will be discussed in later chapters.) Quantum theory and general relativity are significant improvements over Newton's and Maxwell's physics. Nevertheless, as special cases of the former two, the latter two are still abundantly practical.

BLACK HOLES: CHALLENGES IN THE QUEST FOR SAMENESS

But since nature is one and beautiful, one and beautiful should also be the theory that would explain it, physicists think. Hence a TOE must be capable of accurately describing all the phenomena of nature from the microscopic to the macroscopic. Unfortunately, general relativity and quantum theory seem irreconcilable, a very displeasing situation in science. For example, by employing relativity we derive one set of properties for black holes and by combining relativity and quantum theory we get a different set of properties. General relativity and quantum theory don't see eye to eye when attempting to explain black holes, but why?

A black hole is a dense object with gravity so immense that, according to Einstein's general relativity (that first predicted such strange things), within a certain distance from the object, nothing, not even light, can escape. The invisible spherical boundary around a black hole that defines this distance is called the event horizon. It is appropriately named because matter and light can fall through the event horizon but cannot escape—and so whatever events might be occurring within a black hole cannot be seen by us. Consequently, if some kind of civilization existed there, its citizens could see us, but we couldn't see them. Could we ever find out what's really happening inside a black hole (within its event horizon), or will this type of information be a truth, a secret of the universe that will always be hidden from us? Since mythical time of Prometheus scientists have never liked the idea of nature eternally withholding its secrets from us.

To answer this question about black holes we need to first answer another more general and fundamental question: what happens to an object that falls into a black hole? What is its destiny? Specifically, could we still somehow recover all the information about everything that happens to it inside the black hole? If yes, then information is considered preserved. But if not, if after a certain point in time we cannot link any information to that particular object, then all information regarding the fate of that object is thereafter considered lost in the black hole. This is an open question, but how is its resolution related to a theory of everything?

Within the context of general relativity, information is lost forever. But according to quantum theory information should be preserved—as this is a fundamental quantum premise: unlike general relativity, the equations of quantum theory do not produce objects with event horizons that can hide or, as we will see immediately below, even destroy information. Let's analyze the destiny of information in a thought experiment by considering what happens to a book falling into a black hole.

According to relativity, first the book crosses easily the event horizon since the invisible boundary is supposed to be an energetically calm region and nothing special to pass through. But ultimately crushed by the immense gravity the book pieces condense at the infinitely dense point-like center of the black hole. What once was a distinguishable book is now just matter indistinguishable from all other matter already present there. And all information about it is thereafter lost forever.

But quantum theory (at least through the "firewall" approach, which is one of various in the study of black holes) sees the fate of that same book differently: although the event horizon is a highly energetic boundary (in the firewall approach), a wall of fire that burns the book as it goes through, still, some energy is radiated to us from the region just outside the black hole that amazingly contains subtle information about the fate of the book (about the changes it under-goes) inside the black hole. Information therefore is preserved. Now the notion of a firewall violates a fundamental premise of general relativity: the principle of equivalence. This principle partly asserts that an astronaut should feel no dif-ference between free-falling into a black hole and accelerating in empty space (far from black holes). Similarly, an object crossing the event horizon should experience no sudden changes and nothing special, but that's not what happens to the book in the firewall.

These two proposed conclusions clearly clash: if information is lost, quantum theory is fundamentally wrong, but if information is preserved, it is general relativity that is fundamentally flawed. Consequently, depending on one's approach, the properties of black holes are contradictory and the question still persists: is information conserved or lost? This contradiction is known as the information paradox.

A possible reconciliation of these two major theories that claims to leave both intact (at least for now) in their battle for cosmic dominance came through a revolutionary proposal in early 2014 by world-renowned theoretical physicist Stephen Hawking (1942–).[8] His proposal shocked the scientific community: through a new set of arguments his educated guess determined that information *is* preserved, which is an encouraging result on the one hand since information preservation is a fundamental premise of quantum theory. But on the other hand, if we are to save the equivalence principle of relativity, information preserva-tion must hold only if black holes do not exist! Hawking's conclusion is ironic since he has been the leading authority and proponent of black holes for the last forty years. But now he claims that matter falling into a black hole can finally be spewed out as radiation, which, when analyzed, can in principle tell us what happened to that matter inside the black hole. Thus information is preserved—though, he continues, this will in practice be as difficult as weather forecasting. But restoring the principle of information preservation for quantum theory and, at the same time, the principle of equivalence for general relativity comes at a high price. According to Hawking, an event horizon, which is the defining characteristic of black holes, cannot exist, and so neither can black holes (at least in the sense of objects with an event horizon from within, which nothing can escape). This of course remains to be seen. Hawking believes that the final verdict concerning the true properties of black holes will be rendered only after we first reconcile general relativity and quantum theory through a theory of everything—that is, one principle/equation that describes consistently all of nature. The horizon of Thalesian sameness has never been more eventful.

THE SAGE

Thales was regarded as one of the seven sages of the ancient Greek world and the wisest among them. For this he was offered a golden cup, which he respectfully declined by offering it to another of the sages, who offered it to another until the cup was again returned to Thales. He then dedicated it to the god Apollo at Delphi. Thales, a humble man, is credited (among several other Greeks) with the famous aphorism "Know thyself." He was a philosopher, scientist, astronomer, mathematician, politician, even a theologian known best for his belief in hylozoism, that "all things are full of gods."[9]

But he was also a practical man. As an engineer, for example, he aided the army of King Croesus of Lydia in crossing the Halys River by digging a deep trench in the shape of a crescent and diverting its waters. It was a Herculean feat indeed. The waters initially flowed by one side of the army but later diverted; they flowed by the opposite side. Through his observations of the night sky he discovered the stars of the Little Dipper (Ursa Minor), which includes the North Star Polaris, and used them to teach navigation. He also wrote treatises on various calendars such as on the spring and fall equinoxes (which by the modern calendar occur about March 20 and September 22), on the summer and winter solstices (about June 21 and December 21), on the phases of the moon, on solar eclipses, and on the rising and setting of certain stars such as the Pleiades.

While in Egypt Thales is said to have computed the height of a pyramid by first noticing that at a certain time of day his own shadow was as long as his height. He then concluded that the length of the pyramid's shadow at that same time of day was, according to the law of similar triangles, equal to the pyramid's actual height. While on land and through the use of geometry he was able to calculate his distance from a ship at sea. Furthermore, he estimated correctly the angular size of the sun and of the moon relatively to the angular size of their apparent orbit in the sky to be equal to $1/720$.[10] Angular size of an object is the angle created from your eye to two diametrically opposite points on the object. Say the dot, •, represents the eye, letter I, the object, and that they are situated like so, • I, from each other. The angular size of I is the angle depicted in the following geometry: •<I. Today we know that the angular size of both the sun and of the moon is 0.5 degrees (a fact that is true because the sun is much farther away than the moon). Incidentally, had these angular sizes happened to be unequal, their apparent sizes would also have been unequal, and, consequently, a total solar eclipse, during which our view of the sun is completely covered by the presence of a new moon in between (like the one Thales predicted on May 28, 585 BCE), would not have been possible. Now the sun's and moon's angular size of 0.5 degrees divided by 360 degrees, which is the angular size of their apparent orbit around earth, is exactly equal to Thales's estimation, namely, $0.5/360 = 1/720$!

Thales was also known for his weather predictions, a skill proven valuable in teaching his fellow citizens an important lesson about life regarding their negative attitude toward philosophy. In spite of all his knowledge (practical and abstract) and all his wisdom Thales is said to have been poor. And because of his poverty some people criticized philosophy by calling it a useless and impractical way of life. According to one account, "As Thales was studying the stars and looking up . . . he fell into a well. A Thracian servant girl with a sense of humor . . . made fun of him for being so eager to find out what was in the sky that he was not aware of what was in front of him right at his feet."[11] But had the great Dante Alighieri (1265–1321) witnessed the incident he would not, I am certain, have made fun of Thales but would, I am still certain, have responded to the girl by saying:

> The heavens are calling you, and wheel around you,
> Displaying to you their eternal beauties,
> And still your eye is looking on the ground.[12]

Hands-on Thales responded similarly, not in words but through a practical action. "He perceived by studying the sky that there would be a good olive harvest. While it was yet winter and he had some money, he put down deposits on all the olive presses in Miletus [his hometown] and Chios [a neighboring island] for a small sum, paying little because no one bid against him [as it was way too early for anyone to worry about the next harvest that would occur during the next autumn and winter]. When harvest time came and everyone needed the presses right away, he charged whatever he wished and made a good deal of money—thus demonstrating that it is easy for philosophers to get rich if they wish, but that is not what they care about."[13] What they do care about is the rational critique of nature.

One phenomenon that was analyzed rationally was the annual overflow of the waters of the Nile River—an unexpected phenomenon within the context of the generally dry Mediterranean summers when it begins to occur—that puzzled several Greek thinkers including Anaxagoras, Democritus, Herodotus, and Euripides. The occurrence was explained in a naturalistic way first by Thales, who ascribed seasonal northerly winds as its cause that hindered the river from emptying into the Mediterranean Sea and forced its waters to spill over its banks. Ancient Egyptians attributed the flooding to the tears of their mourning goddess Isis over the loss of her husband, Osiris. Today we know that the Nile's overflow is due to seasonal precipitation (mainly rain) on the high-lands of Ethiopia (south of Egypt), where one of the sources of the Nile can be traced. Incidentally, Democritus's explanation was similar to ours.[14] That Thales was wrong is not so important as is his attempt to offer a rational explanation for this natural occurrence. Similarly, Thales and the pre-Socratics in general treated eclipses as natural phenomena, whereas the Babylonians viewed them as omens, despite the fact that the latter kept fairly accurate records for their repeated cycles. Comets, too, were generally thought of as bad omens, but not by the pre-Socratics. Anaxagoras and Democritus, for example, thought that comets "are a conjunction of planets that, when coming near each other, create the illusion that they touch,"[15] an explanation, which although incorrect, is logical because it explains why a comet appears to be a strip-like light in the sky instead of point-like, as planets and stars are. Today we know that the strip-like appearance is due to a comet's tail. It is created from the evaporation of some of its ices, while a comet approaches the sun in its elliptical orbit around it.

Another example of a naturalistic explanation is found in Thales's cos-mology. Although again wrong, it was reached by a rational analysis. According to it, everything is in essence water. Mud, and so, consequently, the earth formed via the solidification of water. Before Thales, the Babylonians had a deceptively similar idea, that nature had once been only water. But unlike Thales, who invoked only natural processes to explain dry land, the Babylonians invoked the supernatural: they imagined their god Marduk creating dry land and thus earth by first placing a carpet over the water and afterward adding mud on top of it—although how the carpet and mud were readily available to the god when nature was only water had not been explained. Naturalistic interpretations of nature were the approach of all the pre-Socratics and remain the approach of modern scientists.

CONCLUSION

By reasoning that all things are ephemeral transformations of *one* primary substance of matter, Thales attempted to attribute an all-encompassing, common, and unifying principle to all the phenomena of nature, the main goal of physicists today, as well as to understand a notion of great importance in science—namely, change. The concept of change (and the degree of change) has been hotly debated for centuries. Some have accepted it as self-evident, and others have flatly denied it as an illusion. Consensus has yet to be found. Every scientist, past and present, has looked to identify a permanent principle in all of the apparent changes. What that principle might be has varied from one scientific theory to another and from one epoch to the next.

Thales was more of a practical man who accepted change undeniably. His student Anaximander was practical but also an abstract thinker. His primary substance of matter was imperceptible and although he, too, accepted change as self-evident, he also required that change in nature obeys laws and happens with measure. But in all the conspicuous changes, he reasoned, something subtle must endure. He called it the *infinite*.

CHAPTER 9

ANAXIMANDER AND THE INFINITE

INTRODUCTION

Anaximander (ca. 610–ca. 540 BCE) taught that the fundamental substance of matter is the *infinite*: a limitless supply of undifferentiated, timeless substance encompassing all the world and manifesting itself as competing antitheses (e.g., hot versus cold). In modern physics energy has properties that are strikingly similar to those of the infinite and so does the much-sought Higgs boson particle of the standard model of physics (nicknamed the God Particle by Nobel laureate Leon Lederman [1929–] for its elusiveness as well as its significance for our understanding of the structure of matter[1]).

THE INFINITE

While itself intangible, the infinite transforms into all concrete things of everyday. Thus it is the true beginning of everything, animate and inanimate. It is also neutral, having no competing opposite. But it transmutes into opposites in struggle with one another—water versus fire, hot versus cold, wet versus dry, light versus darkness, sweet versus sour, and so on. The unjust dominance of one opposite over the other is ephemeral, for eventually it is rectified at annihilation; then, neutralized, both opposites transform again into the neutral infinite. And since eventually the effects of one opposite cancel those of the other, their endless creations and annihilations neither add anything to the infinite nor subtract. Thus even through its transformations, the infinite remains eternally conserved. In modern physics, it is energy that is conserved through its transformations into competing opposites, that of matter and antimatter, and, like the infinite, energy is also limitless and everywhere.

ENERGY AND THE INFINITE

In physics the notion of energy includes mass, too, since, according to Einstein's famous equation $E = mc^2$, from his theory of special relativity, energy (E) and mass (m) are equivalent and transmutable into each other—they are connected via the speed of light (c). Like the infinite, energy is limitless, timeless, indestructible, and omnipresent even in "empty" space (a challenging concept to be revisited in chapter 17, "Democritus and Atoms" when discussing his atoms in the void). Even more, energy causes change by continually transforming from one form to another (e.g., from light to heat) and from pure energy (e.g., light) into matter (e.g., electrons) and antimatter (e.g., antielectrons, also known as positrons). But even with these transformations the total energy content of the universe is always constant. This is known as the law of conservation of energy. One can neither add more energy to the universe nor subtract any from it. Conservation laws, of which there are several in physics, ensure that changes in nature occur with measure, as in the theory of Anaximander. Measure, in modern physics, means that in all of nature's changes some things/properties remain numerically equal (e.g., energy). This equality (this measure) is in basic agreement with the view of Anaximander, who, to save nature (as we will see), reasoned that neither of two opposites could ever dominate totally. Now, to appreciate further the notion of measure and the similarities between energy and the infinite, we need to first understand matter and antimatter since these, in modern physics, are opposites in struggle with each other, created from energy, and into energy once again they return.

Every particle of matter has a corresponding antiparticle of antimatter. A particle (like the negatively charged electron) and its antiparticle (the positron, which is really a positively charged electron) have the same mass and opposite electric charge (of equal magnitude). They are regarded as competing opposites since when they meet they annihilate each other by transforming completely into pure energy—like Anaximander's water and fire that neutralize each other and transform into the infinite. Furthermore, as opposites, not only do they compete—interact via the forces of nature they obey—but, since their effects cancel each other out, they compete with measure, by obeying conservation laws such as the conservation of energy.

For example, an electron and its competing opposite, a positron, can be created out of energy, interact—they initially move apart but after a brief time in existence they recombine—and ultimately annihilate (neutralize, cancel) each other by converting their masses *entirely* back into the energy from which they came. Just like Anaximander's opposites, which are created and annihilated from and into the infinite. Just as the infinite remains constant during such processes, so does the energy since, according to the law of conservation of energy, the energy content of the universe is the same before the creation of the electron-positron pair, during its existence, and after its annihilation. The energy content never changes, only the forms in which it manifests itself.

Other conservation laws are also obeyed, such as that of the electric charge. In this case, the electric charge of the energy from which the pair was created was zero—pure energy always has zero electric charge: the electric charge remains zero when the pair is in existence—for an electron has an electric charge of -1 (in some units) and the positron +1—and continues to be zero when the pair is annihilated as it once again becomes pure energy.

Let us take the conservation of the electric charge one step further. Since the net electric charge is always conserved, neither type of electric charge has an absolute dominance, as Anaximander would require. Nonetheless, one type of charge has a relative dominance over the other. The negative electric charge dominates temporarily at the vicinity of the electron, whereas the positive electric charge dominates temporarily at the vicinity of the positron. It is this temporary dominance that creates the electromagnetic force of interaction and overall competition between opposite charges. It is the cause of the opposites' becoming and decaying, their attractions, repulsions, motions, conversions to and from energy, and in general, such temporary dominance is a contributing cause of the phenomena of nature.

Competing opposites are necessary for Anaximander and modern physics, as they will also be for Heraclitus, if nature is to remain diverse, eventful, and beautiful. There is an electron here but a positron there. They, and other particles and antiparticles from similar processes, convert to light and to heat. The particles form atoms, molecules, composite objects like the sea, the trees, the breeze, the earth, the sky, and forms of life. It is summer here but winter there. It is warm now but will be cool later, night now but was day earlier. The unity of the world is preserved in harmony by the very competition between opposites. Temporary dominance and the resulting struggle of opposites produce the rich plethora of diverse phenomena while simultaneously, cosmically (universally) absolute dominance is not, *should not* be allowed, for conservation laws must be obeyed. Not only does Anaximander's worldview see the nature of nature as being cosmically just, but because of conservation laws so should modern physics. Curiously, however, it appears that nature is not cosmically just, for one of the most puzzling questions of science today is why there is more matter than antimatter in the observable universe, a question to be pondered in a section below after we first elaborate on the notions of opposites and neutrality a bit more.

NOT AN ORDINARY THING

Anaximander reasoned that the primary substance of the universe could not have been any one of the ordinary things, such as water or fire. For they have opposition with one another, and opposites destroy; they do not generate one another. If water were the infinite—that is, if everything in the universe were initially water—it would be impossible to have its opposite, fire, ever created, for water destroys fire; it does not generate it. And that would be terrible because in such a scenario eventful beautiful diversity would be absent from the cosmos. Thus something that has an opposite cannot be the primary substance of the universe for it presents a serious threat to the cosmic justice, to the unity and order of nature, to its diversity and in fact, to the very existence of nature itself—for such type of substance, with an opposite, would cancel itself out, and thus it would cancel nature itself!

Anaximander saved the phenomena—kept nature eternal, diverse, and eventful, and without the possibility of absolute dominance by any one of the opposites—by requiring that the primary substance, the infinite, be neutral—with no competing opposite. It must be neutral to itself and to the opposites that it creates. With such choice, neither opposite is a threat to nature any longer, since their effects cancel each other out, nor is the infinite, since it has no competing opposite to cancel out. Hence, unlike the opposites, the infinite is permanent and indestructible. And so then is nature itself, for nature's essence is the infinite. The neutrality of the infinite saves the phenomena, but the opposition of the opposites beautifies them. Both neutrality and opposition are central ideas in the world outlook of Anaximander and of modern physics.

The modern physicist's version of Anaximander's reasoning would be that the presently accepted primary particles of matter, the quarks and leptons, cannot really be primary, for they have opposites, the antiparticles of antimatter, the antiquarks and antileptons, and as opposites, a particle and its antiparticle annihilate, not generate, each other. Furthermore, there are six types of quarks and six types of leptons, and the pressing question is why there are so many building blocks of matter and why do they have different general characteristics (different electric charge, mass, spin, etc.)? Why not just have one primary substance, one infinite-like type particle?

On this issue, Nobel laureate Werner Heisenberg (1901–1976) has said, "All different elementary particles [particles that are not made of other things, thus have no substructure, such as the quarks and leptons] could be reduced to some universal substance which we may call energy or matter, but none of the different particles could be preferred to the others as being more fundamental. The latter view of course corresponds to the doctrine of Anaximander, and I am convinced that in modern physics this view is the correct one."[2]

We have seen how energy may be regarded as the infinite, but what kind of material particle of modern physics has infinite-like properties, including the key property of neutrality?

THE GOD PARTICLE

The search for a universal *neutral* substance that addresses Anaximander's concern and saves the phenomena has never been more intense. The particle with the most qualities required by such a substance is the Higgs boson. It has been mathematically predicted to exist by various physicists in the 1960s, including Peter Higgs (1929–), from whom it took its official name. In fact it is required to exist in order to save the phenomena as described by the standard model of physics: the Higgs particles are thought to give mass to all material particles (such as quarks and leptons) by pulling on them, thus forcing them to slow down, clump, and form all the composite objects in the universe, from nuclei, atoms, molecules, plants, animals, planets, and stars, to galaxies, and in general, all the complexity in the universe. The mass-giving mechanism of the Higgs and whether mass is truly an intrinsic property of particles or an acquired one through their inter-actions are more relevant topics for chapter 17. With the Higgs we can explain mass, and as a result the universe is diverse, beautiful, and saved. Without the Higgs, all particles would be massless, would fly around at light speed, and would not be able to come together and form atoms or in general composite objects, including us; the universe in such a case would exist in one boring, undiversified state, which of course would be in contradiction to the actual diversified universe we live in. Indeed, then, the Higgs saves the phenomena!

Like the infinite, the Higgs particle is intangible, neutral (in a few inter-esting ways, as we will explore in the next section), and the field that represents it permeates all of space. (Use of the term "field" means that something exists everywhere, whereas "particle" implies that something exists only somewhere. In quantum theory particles are manifestations of field fluctuations. Consider this analogy: if the sea is a field, a splash in the sea—a fluctuation, an excita-tion—is a particle that can be detected. Because of its all-pervasiveness, the Higgs field was related to Anaximander's infinite by Lederman.[3]) But while itself neutral, the Higgs also manifests itself as the competing opposites, as par-ticles of matter and antiparticles of antimatter, and as some of the forces that matter and antimatter obey; in fact, it is not expected to be observed directly (analogously, neither was the infinite), instead its existence was confirmed indi-rectly by studying the behavior of various other particles that the Higgs decays into. Thus, the observed opposites in nature are in a sense different aspects of the same thing, the Higgs—or, analogously, the infinite. Anaximander proposed the infinite to put his opposites, and Peter Higgs (by modifying the standard model) proposed the Higgs field to put his opposites (the particles and antipar-ticles) in order to explain why they have mass.

Michelangelo's simile of a formed, asymmetric statue inside an unformed block of marble awaiting to be exposed by the sculptor can be applied to our discussion of the infinite and the Higgs. First, note that the unformed block of marble is symmetric hence boringly uniform. To the contrary, the formed statue is asymmetric hence beautifully irregular, with curves, lines, expression, and character. With this in mind, the asymmetric opposites (that is, formed matter, fire, water—the beautiful statue) are hidden in the unformed, symmetric nature—the dull but necessary raw marble—until the sculptor of nature, the infinite or the Higgs (or in general, the human mind), breaks her symmetry and exposes them.

Now this simile can be discussed within the context of a more literal scenario, that of the cosmological model known as the big bang, which speculates how the universe got started and how it has been evolving. According to this model, about 13.8 billion years ago the entire universe was unimaginably small, possibly a mere point, infinitely dense and hot. It then exploded in the absolutely most extraordinary event called the big bang and has ever since been expanding, cooling, rarefying, and creating the eventful universe we live in. Thus the universe was once, and for a minuscule moment during its early blazing stages of existence, much smaller, denser, hotter, perfectly (or more) symmetric than today (i.e., with all forces unified), and with none of the particles having any mass. An inconceivably small fraction of a second after the big bang, when the universe expanded and cooled, the Higgs field was activated, broke nature's initial symmetry by giving mass to only some particles, and has ever since been playing a crucial role in the plethora of beautiful and diverse phenomena we observe today. Analogously, Anaximander hypothesized that nature was once uninterestingly symmetric and unstructured until the infinite's eternal motion caused the eventful opposites to separate out from it and break the monotonous symmetry.

NEUTRALITY

To save the phenomena, neutrality must be an essential characteristic of a primary substance. The Higgs is neutral in various ways: (1) it is electrically neutral, thus it is its own antiparticle—it is both matter and antimatter, and in this respect it has no warring opposite to be destroyed by (note, even when the Higgs particles decay, the Higgs field still endures and is all-pervasive). (2) It is also color-neutral— "color" (or more precisely, color charge) here is a property of the quarks and of the nuclear strong force (like the electric charge that is a property of the electromagnetic force) and not the color of everyday sense. Parenthetically, color charge comes in three flavors: blue, green, and red. These color charges represent values that control the strength of the nuclear strong force between quarks. Where do their colorful names come from? The electric charge, which produces the electromagnetic force, has two values: the positive and negative electric charge. The color charge, which produces the nuclear strong force, has three values, which could have been called anything, such as A, B, C. But, because physicists are interesting people, they were fancifully named after the three primary colors of light, so blue, green, and red. (3) Moreover, its spin is zero, thus it is direction-neutral, an unusual notion that we need to elaborate on further.

Spin (like electric charge) is an intrinsic quantum property of elementary particles. In a simplistic view, imagine the spin of a particle to be like the spin of a top around its axis. However, unlike a spinning top, which may spin slow or fast and in every direction, an elementary particle spins with a fixed magnitude and only in certain directions. Now, the direction of a particle's spin is related to the direction of its motion through space. For example, a neutrino (an electrically neutral point-like particle, belonging in the family of leptons as electrons do) is observed to always be left-handed. This means that a neutrino moves through space like a left-handed screw: it advances (moves forward) by spinning counterclockwise. An antineutrino on the other hand is always observed to be right-handed. It moves like a right-handed screw: it advances (moves forward) by spinning clockwise. Unlike a neutrino, an electron can be ambidextrous—or move in space by spinning in either direction. The important point is that all particles are directional—the direction of their motion through space is restricted by how they spin.

But a primary substance must itself be free of such restriction; it must be nondirectional, direction-neutral, that is, isotropic (with no preferred direction of motion)—because a left-handed substance would, in Anaximander's terms be like, say, fire, and a right-handed substance like water. If a primary substance's own motion were restricted it would not have been able to generate the existing particles with all their various observed directionalities, which collectively are isotropic. In other words, if Anaximander's cosmic justice is to hold, a preferred (special) direction in the universe, toward which particles would be moving, should not exist. In fact, *on a grand scale*, the view of the universe is similar in all directions, thus the universe itself is isotropic (there is no special direction in it). One proof of this is the observation of the so-called cosmic microwave background, light that comes to us from every direction in the universe, showing that the matter that emitted it had almost exactly the same temperature, about 3 degrees above absolute zero.[4] Now, since the universe is made of galaxies and stars, which in essence are made of particles, then the isotropy of the universe must really be a consequence of the isotropy of its constituent particles. And if we believe that one day we will conceive a theory of everything, describing one primary substance, such a substance must have the property of isotropy, for only then it can generate the universal isotropy that we observe today. Well, to be isotropic, direction-neutral, the spin of such primary a substance must be zero since without spin the direction of its motion cannot be restricted. The Higgs boson particle is electrically neutral as well as color-neutral, and being the only particle of the standard model with zero spin, it is direction-neutral, too! Anaximander's notion of neutrality should be a vital property of the primary substance of the universe, and consequently of the universe itself. Nonetheless, it seems that the universe does not obey this fundamental notion of neutrality! For although it is isotropic and thus cosmically just (neutral) direction-wise, it is also unjust matter-wise; matter appears to dominate antimatter! What happened to the cosmic justice?

WHY IS THERE MORE MATTER THAN ANTIMATTER?

In modern cosmology, there is an open question: why is there more matter than antimatter in the observable universe? In Anaximander's terms, this problem might have been phrased: Why is there more water than fire in the observable universe? This observation makes no sense if indeed the universal substance is a kind of neutral, which transforms into equal amounts of opposites with properties that cancel each other out through the conservation laws they obey—so that the universal substance can remain neutral. Absolute dominance by any one opposite should not be allowed. Yet matter appears to have an absolute dominance in the universe. If true, and our observations are correct, where is Anaximander's cosmic justice? Are the laws of physics as we now know them really incorrect? The answer does not yet exist, but I will speculate cautiously.

Think of a creation process like that of an electron-positron pair. It obeys cosmic justice. During its existence, the region where the electron is located is dominated by matter, and the region where the positron is located is dominated by antimatter. But such dominance is relative and ephemeral. No law forbids equal and relative dominance by matter and antimatter in various regions of the universe. In fact this would be expected if indeed equal amounts of matter and antimatter are generated by a neutral-type universal substance. What *is* forbidden is *absolute* dominance, for which the amount of either matter or antimatter in the universe is absolutely more. Now, with relative dominance in mind we can speculate on why there is, or actually, why there *appears* to be more matter than antimatter in the observable universe.

First, the universe is immense, and we haven't observed all of it yet. So an unseen part of it might be composed of mostly antimatter that could balance out the matter we see, hence we have cosmic justice.

Second, what if our universe is not *the* universe but rather is only but a mere universe, a mere region, in a multiverse (which, by definition, is supposed to be composed of many universes, perhaps like ours, perhaps different)? In fact, we should not be rash in dismissing such a view, for not too long ago we thought the entire universe was what we now call the Milky Way galaxy—but we now think there are some 170 billion galaxies. If the multiverse hypothesis holds, the puzzle of the observed asymmetry between matter and antimatter might then be resolved. Matter might be dominating temporarily in our universe, but antimatter could be dominating temporarily in another universe, and in such a way that neither can claim absolute dominance in the world—the multiverse. Because such temporary dominance can be neutralized when such universes collide, converting their matter and antimatter to pure neutral energy. Anaximander's cosmic justice would then be restored.

Anaximander believed that there exist innumerable worlds, coming into being from the infinite and perishing back to it. Analogously, there are many modern cosmological models that speculate on many universes, all of which are part of a multiverse. Such hypotheses are a result, for example, of various interpretations of quantum theory, such as the so-called many-worlds interpretation (to be introduced in chapter 16, "Anaxagoras and Nous"), or attempts to construct a viable theory of everything.

COSMOLOGY

In addition to the hypothesis of the abstract infinite, Anaximander makes another conceptual leap by holding that the earth is motionless in space and without any physical support. This happens, he argues, because of its equal distance from everything, which is also uniformly distributed around it, and because of its equilibrium (its equal tendency to move in every direction).[5] This is in contrast with Thales's view of an earth floating on water and thus supported by it. Philosopher Karl Popper (1902–1994) remarked, "In my opinion this idea of Anaximander's is one of the boldest, most revolutionary, and most portentous ideas in the whole history of human thought. It made possible the theories of Aristarchus and Copernicus. But the step taken by Anaximander was even more difficult and audacious than the one taken by Aristarchus and Copernicus. To envisage the earth as freely poised in mid-space, and to say 'that it remains motionless because of its equidistance or equilibrium' (as Aristotle paraphrases Anaximander), is to anticipate to some extent even Newton's idea of immaterial and invisible gravitational forces."[6] Aristarchus of Samos (310–ca. 230 BCE), sometimes called the ancient Copernicus, was the first to advance the heliocentric model, revived in the sixteenth century by Copernicus (perhaps Copernicus could be viewed as the modern Aristarchus).

As commended by classicist John Burnet (1863–1928)[7] Anaximander's doctrine of innumerable worlds, each with its own earth, heaven, planets, stars, and especially its own center and diurnal rotation, is inconsistent with the existence of an absolute center or preferred direction of motion in the universe. With the lack of an absolute direction of motion, Burnet continues, Anaximander's argument that an earth that happened to be equidistant from everything in its world has no reason to move in any direction is quite sound. This is in fact a clever use of symmetry, a notion of central significance in modern physics. Symmetry, as in an unformed block of marble or a circle, implies a certain constancy, a similarity that persists without change. The unformed marble is the same throughout, for example, or the circle looks the same from its center at all angles, or, in Anaximander's case, the earth remains in equilibrium because of its equidistance. Symmetry in physics does not describe just appearance. It also underlies conserved abstract properties of nature such as the conservation of energy, momentum, and electric charge. For example, conservation of energy is a consequence of the hypothesis that the laws of nature are symmetric (invariant) with respect to time translations: they work the same today as they have in the past and are expected to continue so tomorrow. Because this hypothesis has been true so far, we accept conservation of energy as a law of nature.

The absence of a special direction in space was employed by Democritus in describing the motion of his atoms as random (see chapter 17). Though true, abandoning the notion of an absolute direction is still difficult today. We are tricked by the phenomena (such as falling objects or "the earth being under our feet and the sky up above us"), and so we often think of up above and down below as if they were really absolute up and absolute down. We don't realize that for those living on the opposite side of the earth, our relative up is really their relative down, and our relative down is really their relative up.

ON LIFE AND EVOLUTION

Thales, Anaximander, and (as we will see in the next chapter) Anaximenes (collectively known as the Milesian philosophers since they were born in Miletus, a Greek city in Asia Minor) explained nature in terms of the variations of *one* universal substance. Nature, of course, includes us and all life-forms. So an immediate consequence of their monistic theories is either (a) that there is no lifeless matter, to the contrary, everything is somehow alive; each philosopher's primary substance (the water, the infinite, or, in Anaximenes's case, the air) is somehow alive and so is everything that it transforms into. Or (b) that humans as well as all other species originated somehow from lifeless matter (the water, the infinite, or the air).

View (a), which is known as hylozoism, was held by Thales and Anaximenes, and although highly controversial, it is still an interesting notion, for, despite 2,600 years of advancements in science and philosophy, a clear-cut distinction between animate and inanimate matter cannot be made. An unambiguous definition of what is alive or dead does not exist as argued, for example, by four Nobel laureates: Charles Sherrington (1857–1952),[8] Erwin Schrödinger,[9] Werner Heisenberg,[10] and Richard Feynman (1918–1988).[11]

View (b) has a certain similarity with the premise of the modern theory of biological evolution—that regards the various species to have evolved gradually from a common ancestor (or two, possibly more) speculated to have arisen spontaneously from lifeless matter. By spontaneously I mean that the exact mechanism of life's origin is not yet known, although chemical reactions are generally the assumed cause. Now, compared to the other Milesians, Anaximander had a more concrete and extraordinary theory of the origin and evolution of the species, including humans, that captures four specific aspects of the modern theory of biological evolution: (1) life arose spontaneously from lifeless matter, (2) more complex life did not arise spontaneously but evolved from the less complex, (3) life's adaptation to its environment, and (4) survival of the fittest.[12]

Equally importantly, his theory was based on an accurately analyzed *observation*. Noticing that human babies are helpless at birth and for several years thereafter, Anaximander argued that humans could not have originated with the young of the species in their present form because they would have never survived. While newborns of other animals quickly support themselves, human babies cannot survive without long parental care. Therefore, he held that humans (and in general all animals) evolved from species, precisely fish, whose newborns were more self-reliant than human (or land-animal) babies.[13]

His general doctrine was that the most primitive forms of life were generated spontaneously in the moist element as it was evaporated by the sun—note that his notion of antithesis is present here, too, as the wetness of moisture versus the dryness of the sun. These living creatures had a protective spiny membrane and were the first kind of fish. With time, he speculated, they evolved to various other forms of fish. Then, some of their descendants abandoned the liquid element and moved to dry land, adapting to different conditions and evolving to new forms of life, including humans.[14] Modern theories of biological evolution are quite similar: primitive microscopic life is speculated to have appeared spontaneously initially in water, evolved to fish, then to the sea-land transitional amphibia, to mammals, to primates, then to the first hominids from which modern humans ultimately evolved.

Newborn self-reliance was the first state in the development of life. Newborn helplessness and long parental care have developed afterward and most probably simultaneously: as a newborn need arose an able parent addressed it. As if the evolution of a bad characteristic happens simultaneously with the evolution of a good characteristic, a notion, which if true, resonates well with Anaximander's doctrine of the simultaneous appearance of opposites: an inefficiency in some area, say, the inability to walk at birth and for several months subsequently, may evolve simultaneously with a corresponding efficiency in some other area, for example, an advanced brain, so that one may moderate the other—for example, the evolution of an empathetic brain allows parents to care for their helpless offspring for a long period. In general, the growing-up time increased with increasing brain size and complexity.

The selfless act of long parenting not only guaranteed the survival of the species but also, I believe, contributed to the overall bond between all people. Because once we began caring for our babies, we gradually began to care for our immediate and extended family, consequently increasing the chances to care more for our village, city, country, the human race, and ultimately life in general.

CONCLUSION

Anaximander's intellectual leap is marked by three of his theories: in cosmology of an earth motionless in space without the need of a physical support, in biology on the origin and evolution of living creatures including the human species, and certainly of his theory on the primary substance of matter. With the latter, he modeled change and diversity in terms of constant transformations of the intangible, neutral, and conserved infinite, into the concrete, competing, and transient opposites of everyday experience, and back and forth and with measure. Nonetheless, it was Anaximenes who formulated the first graspable theory of change—of how matter can transform between its various phases: the gas, the liquid, and the solid.

CHAPTER 10

ANAXIMENES AND DENSITY

INTRODUCTION

I n his search for the primary substance of matter, Anaximenes (who flourished ca. 545 BCE) returned to the tangible world and chose air. His way of
studying nature was economical and straightforward. Starting with a single
material (air) of unchangeable nature, he managed to explain the manifold of
natural phenomena quantitatively, in terms of condensation and rarefaction of
matter. For with these opposite processes in mind, it was no longer necessary
to ascribe all sorts of different properties to each object—such as rigidity, softness, hotness, coldness, wetness, dryness, fluidity, weight, color—just how dense
it was. This idea in itself has a certain truth. But from a grander point of view as
regards the evolution of science, this idea had surely been catalytic for the discovery of one of the most successful theories in science: the atomic theory.

CONDENSATION AND RAREFACTION

In Greek, "air" refers to any gas, and quite possibly in Anaximenes's view air was
vapor water. His main question, however, was how a single material, air, in its
gaseous state, could be transformed into all other forms of matter and account
for the overabundance of dissimilar things, while itself remaining unchanged.
What mechanism or processes could be applied to air, keep its substance
unchanged, yet convert air into all the different things—solids, liquids, and
gases? Change, he proposed, occurs via two opposite processes: condensation
and rarefaction of matter.[1] Successive condensations of gases transform them to
increasingly denser matter, the liquids and solids, but successive rarefactions of
solids transform them to increasingly rarefied matter and once more back to the

liquids and gases, an essentially accurate idea. These processes cause changes in the density of matter but do not alter the very nature of matter (its very substance). Hence, every object is really air—in general, made of the same material—condensed or rarefied.

WHY AIR?

Air is in various ways of simpler form than other everyday substances. It is highly mobile and can be found almost everywhere. It is invisible, thus apparently unstructured and symmetric, rarefied, thus quantitatively less. Symmetry, perceptible or subtle, was and still is a much-desired characteristic for nature in both ancient and modern scientific theories. In addition, starting from less (at least in a quantitative and visual sense, e.g., rarefied invisible air) and aiming to explain more (e.g., a denser, thus quantitatively more, visibly more structured, and thus in a way more complex substance), has always been the preferred approach in both science and mathematics; in mathematics, the fewer the assumptions (axioms), the more powerful a theorem is. Parenthetically, as regards religion, the reverse is true: polytheism preceded monotheism.

Now, although Anaximenes thought that fire is rarefied air and thus quantitatively less than air, still, as a primary substance fire seems to not have been adequate for him, for unlike air, fire is visible, has a variable form, and so is structured and asymmetric. Furthermore, air is needed for life through breathing, whereas fire destroys life. In fact air's traditional association with soul (from the pre-Homeric times) might have influenced Anaximenes, for he writes: "Just as our soul, being air, holds us together, so do breath and air encompass the whole world."[2] The significance of fire in the explanation of natural phenomena will be elevated in the philosophy of Heraclitus (see chapter 12).

Lastly, Anaximenes was an empiricist, thus he abstracted his theory as a consequence of careful observations of various meteorological phenomena for which air had (or so he thought, anyway) a significant role. "When it [air] is dilated so as to be rarer [more rarefied], it becomes fire; while winds, on the other hand, are condensed air. Cloud is formed from air by felting [due to condensation]; and this, still further condensed, becomes water. Water, condensed still more, turns to earth; and when condensed as much as it can be, to stones."[3] He imagined objects to be in either one distinct phase (the solid, liquid, or gas) or in a mixture of phases; "Hail is produced when water freezes in falling; snow, when there is some air imprisoned in the water."[4]

FROM RAREFACTION AND CONDENSATION TO THE ATOMIC THEORY OF MATTER

It will be argued that the discovery of the ancient atomic theory of Leucippus and Democritus—of atoms in the void—followed as a logical consequence of the ideas of rarefaction and condensation. And as we will see in chapter 17, modern atomic science has its roots in the atomic theory of antiquity.[5]

Softness and the Void, Rigidity and the Atoms

Softness occurs with rarefaction and rigidity with condensation, Anaximenes held, but how? Since everything is made of soft and penetrable air, why are some objects (e.g., the solids) rigid and impenetrable? Why is a piece of metal (which is supposed to be condensed air) incompressible and impenetrable, while air is compressible and penetrable? Why can we walk through air (so it seems, anyway) but not through a solid wall (which is supposed to also be air)? How do rarefaction and condensation really work, and how can we explain the varying degree of softness or rigidity in an object? Furthermore, what keeps matter together in a condensed or rarefied state?

First, let us take all four notions for granted—rarefaction, condensation, and the resulting softness and rigidity in an object. Then we ask: If we could imagine rarefaction and condensation to occur ad infinitum, what kind of an object would the absolutely most rarefied or condensed be? The most rarefied type of object would have zero density and would be *absolutely* soft, compressible, and penetrable; as if the object were void of matter; as if it were immaterial and did not exist; as if it were nothing! Now, an object void of matter is really a *void*, empty space. And so the most rarefied type of objects could be thought of as material-less gaps in space. On the other end of the limit, the most condensed type of objects would still be of the *same* substance, would have infinite density, would be *absolutely* rigid, incompressible, and impenetrable, and could be thought of as the matter that is filling up the nonempty space. These impenetrable pieces of matter that are also disconnected from each other by the void between them, are precisely the atoms of Leucippus and Democritus, and the philosophically controversial void is precisely what they invented to facilitate the motion of their atoms.

"There are but atoms and the void," said Democritus,[6] or equivalently, "the full"[7] and solid, and "the empty"[8] and rarefied. First note that Anaximander's opposites are present here, too, for the full and solid is the opposite of the empty and rarefied. Furthermore, "the full" seems to correspond to the aforesaid absolutely most condensed object, and "the empty" to the absolutely most rarefied object. Comparisons and details of the ancient and modern atomic theories will be carried out in chapter 17. For now it suffices to visualize tiny, indestructible atoms, absolutely solid—and unable to rarefy—of all sorts of shapes moving randomly through the void, colliding with each other, either hooking with one another and clustering (condensation), or unhooking and dispersing (rarefaction). Thus with atoms moving in the void, we understand how condensation and rarefaction are actually carried out—how matter can move, assemble, and stay together, or disassemble. And the consequent rigidity or softness is determined by the density of an object, that is, by how many atoms are cramped together within an object and by how much void is between them through which to move: the less the void, the more rigid the object; the more the void, the softer the object. Could the ancient atomic theory have been discovered through such type of analysis?

I don't know if the atomists, Leucippus and Democritus, discovered atomism by analyzing rarefaction and condensation in the two extreme limits just described, but they were certainly capable of doing so, especially the great geometer Democritus. For such type of thinking, which in mathematics is part of what is known as the theory of limits, had already been invented and applied by him in other cases (e.g., in the calculation of the volume of a cone). In fact Democritus's knowledge on limits was commended by the great astronomer Carl Sagan (1934–1996) in this quote: "Perhaps if Democritus' work had not been almost completely destroyed, there would have been calculus by the time of Christ."[9] By "work," of course, Sagan meant, among other things, Democritus's knowledge of limits, for to invent calculus a prerequisite knowledge is the theory of limits. Calculus was finally invented independently by Isaac Newton and Gottfried Leibniz (1646–1716) in the late seventeenth century.

One other way that the mathematical analysis of rarefaction and condensation might have aided the discovery of atomism is discussed below.

Continuous versus Atomic

The challenge to understand how condensation and rarefaction themselves are carried out was important for the evolution of scientific ideas because it forced Anaximenes's successors to think profoundly about the nature of matter. Consequently, they discovered two antithetical views: the continuous and the atomic (the discontinuous). The complexities associated with the former guided the mathematical genius Democritus, the last of the pre-Socratics, to atomism. Erwin Schrödinger argued that the mathematical challenges of the continuum were related to similar challenges of a continuously distributed model of matter.[10] For example, we cannot tell how many points a purely mathematical line has. Analogously, if we have a material line (or in general an object), we cannot tell how many material points it has and how these points behave during rarefaction and condensation. Namely, how can an unchangeable substance of matter (e.g., Anaximenes's air), distributed *continuously* within an object, rarefy or condense? "What should recede from what [so that an object can rarefy, or what should approach what so that an object can condense]? . . . if it is a *material* line and you begin to stretch it—would not its points recede from each other and leave gaps between them? For the stretching cannot *produce* new points and the same points cannot go to cover a greater interval."[11]

In other words, can matter, modeled as continuously distributed in space, really move through other matter in order to condense or rarefy? Can new matter move into and occupy the space that is already occupied by other matter? When matter moves, where does it move into, and what does it leave behind? How do condensation and rarefaction really work if matter is continuous? They do not! They work only if matter is discontinuous: made of disconnected, indivisible, and incompressible pieces—the atoms of Leucippus and Democritus—

moving in the void. Rarefaction occurs when the atoms in an object recede in the empty space around them, and condensation occurs when they come closer to each other.

Modeling matter as discontinuous (atomic) constituted the very first quantum theory, the precursor of the modern. In modern quantum theory both matter and energy are quantized (discontinuous): matter is composed of disconnected elementary particles, the quarks and leptons, and energy comes in discrete (quantum) bundles (e.g., photons are the particles of the energy of light).

CONCLUSION

All challenges of rarefaction and condensation could be accounted for only through atomism (to be introduced fully in chapter 17), since such a great idea required first the development of all other great ideas conceived by Democritus's predecessors, but also in the light of mathematics (Democritus, the principal contributor of the atomic theory, was a brilliant mathematician). The significance of mathematics, not just as an abstract field of knowledge but also as a practical method to describe nature, had been realized early on, especially by the great Pythagoras as a consequence of his passion for numbers.

CHAPTER 11

PYTHAGORAS AND NUMBERS

INTRODUCTION

Pythagoras (ca. 570–ca. 495 BCE), a pioneer in applying mathematics to the investigation of physical phenomena, consequently initiated the mathematical analysis of nature, a cornerstone practice in modern theoretical physics. "Things are numbers" is the most significant Pythagorean doctrine.[1] While its exact meaning is ambiguous, it probably signifies that the phenomena of nature are describable by equations and numbers. Therefore, nature is quantifiable. That is, properties such as hotness, brightness, loudness, wetness, softness, and in fact all characteristics of nature, are measurable. Based on this, the underlying principle of nature is not material (e.g., water, air) but is rather a mathematical form (an equation). Since the mathematical representation of nature is not readily realized, the doctrine was emphasizing that sense perception was merely revealing an untrue version of nature (reality), a truer version of which could be glimpsed by the intellect through modeling nature mathematically. How one begins to model nature mathematically will be discovered in this chapter via the work of Pythagoras and his students. They quantified pleasing sounds of music, right-angled triangles, even the motion of the heavenly bodies.

THE MAN

Pythagoras founded a school in Croton in southern Italy, open to both men and women, where he and his students pursued various studies, including religion, science, philosophy, mathematics, and music. They practiced a common way of life: asceticism (through body exercise, a vow of silence, a special diet that

135

avoided meat and fish) and secrecy (probably to keep their discoveries exclusive to their students in order to attract more members to their school). Therefore, it has always been difficult to distinguish exactly what philosophical views belong to him or to some other Pythagorean. Aristotle avoids such difficulty by often referring generally to "the Pythagoreans." Plato in *The Republic* writes specifically about Pythagoras and says that he was uniquely respected and loved by his students, not only for his knowledge but also for teaching them "the Pythagorean" way of life, best known for its high ethical standards. Wisdom, justice, and courage were among the sought-after virtues. Friendship was also highly valued. Pythagoras's dogmatic influence on his students was evident by their reference to his opinions as prophesies with the characteristic phrase "He himself said so."[2]

COSMIC HARMONIES

Pythagoras's first application illustrating the role of numbers in nature was the mathematical description of mellifluous sounds of music. First, he discovered that in stringed instruments the sound of a plucked string depends on its length and tension. For example, the sound of a plucked guitar string is of a higher pitch as it is pressed down and made shorter by your finger. Then he observed that the blended sound produced by two plucked strings of the same tension is more pleasing when their lengths are in ratios of small integers—for example, 2:1 is the octave, 3:2 the fifth, 4:3 the fourth, and 5:4 the third—thus numbers forming a discrete, a *quantum* set—which for the example given is the set of 1, 2, 3, 4, and 5 (obtained by arranging in sequence the numbers of the above ratios).

The Pythagorean theory of music was a significant milestone in the evolution of science from two points of view. First, since the phenomenon of sound can be quantified, that is, it is represented by mathematical formulas, why not all phenomena? Second, if all things truly can be represented by numbers, then mathematics is the underlying and unifying principle of every natural phenomenon, even the seemingly dissimilar. With this in mind, everything may somehow be related at least mathematically—in other words, a kind of master mathematical equation may be invented that can describe everything. So phenomena that have no apparent relationship with one another, on a deeper level, mathematically, may prove to obey the same mathematical principle and thus have something subtle in common. We have already spoken about unification efforts undertaken nowadays in search of a theory of everything. But the first step of cosmic unification in search of a universal law was taken with the intellectually bold Pythagoreans when they connected mathematically two seemingly unrelated phenomena: their earthly harmonies with the heavenly motions! How did they do this?

First they supposed that, similar to the way an object on earth moving through air can produce a sound (slow movement making a low pitch, fast movement a high one), the stars (including the sun), moon, and planets (including earth), moving through ether (the purer air believed then to fill the universe) can produce their heavenly sounds. But these sounds must blend into a song harmoniously. They reasoned that the ratios of the length of strings that produce the harmonious sounds in string instruments *must be the same* as the various ratios formed by the speeds of the revolving heavenly bodies. This requirement restricts basically the speeds and orbits of the heavenly objects to certain discrete, quantum numbers, an idea that resonates with the essence of modern quantum theory!

Relative speeds of heavenly bodies could be easily deduced by comparing each body's rising time. For example, the moon rises about fifty minutes after the stars (some reference group of them that on some day rises together with the moon), but the sun rises only about four minutes after the stars, well-known facts in antiquity. With this in mind, the apparent revolution speed of the stars is the fastest, of the sun the second fastest, and of the moon the slowest. Since their speeds were different, the sounds they would produce as they move through ether would also be different, to say the least. But they also had to be harmonious, Pythagoras conjectured, since nature was a "*cosmos*,"[3] a term credited to him, a beautiful and well-ordered universe for which a cacophonous music of heavens was unaesthetic. The music of the heavenly bodies is inaudible, the Pythagoreans explained (as Aristotle tells us), because it is continuously playing and "the sound is in our ears since our birth, thus it is indistinguishable from its opposite silence; sound and silence are distinguishable only via their mutual contrast."[4] In a parallel example, a cook does not smell his own food after a few hours of cooking it. The earthly string harmonies, which could be heard, inspired the Pythagoreans to deduce by analogy the heavenly harmonies, which could not be heard. This type of approach, to come up with a general law by analogy of something specific, is common practice in science.

Such unusual interconnection was celebrated first in 1619 with the harmonic law of Johannes Kepler (1571–1630) when the astronomer discovered that, as planets revolve around the sun in their elliptical orbits, the ratios formed by each planet's fastest speed at perihelion (a planet's closest distance from the sun) over its slowest at aphelion (its greatest distance from the sun), are very close to the Pythagorean ratios of pleasing harmonies in stringed instruments. In his book *The Harmonies of the World*, Kepler wrote, "The heavenly motions are nothing but a continuous song for several voices, to be perceived by the intellect, not by the ear."[5]

Moreover, in the beginning of the twentieth century, the seminal era of quantum theory, physicists Niels Bohr (1885–1962) and Arnold Sommerfeld (1868–1951) conceptualized the atom as a miniature solar system, with the electrons orbiting the nucleus of an atom like the planets are orbiting the sun. In their theory the orbits of the electrons are restricted to certain discrete speeds and sizes (as were the heavenly bodies in the Pythagorean theory) that are expressible in terms of specific integers called quantum numbers that "display a greater harmonic consonance than even the stars in the Pythagorean music of the spheres [heavenly bodies]."[6] Remarkably, unlike the Pythagorean theory of planetary motion, which was quantized, the Newtonian theory was not: planets, according to Newton's theory of gravity, do not have a restriction in their speeds or orbital sizes. But they should, according to quantum theory, although their quantum behavior is negligibly small because of their large mass.

Even more so, according to the latest developments in string theory and in the words of string theorist physicist Brian Greene (1963–), "everything in the universe, from the tiniest particle to the most distant star is made from one kind of ingredient—unimaginably small vibrating strands of energy called strings. Just as the strings of a cello can give rise to a rich variety of musical notes, the tiny strings in string theory vibrate in a multitude of different ways making up all the constituents of nature. In other words, the universe is like a grand cosmic symphony resonating with all the various notes these tiny vibrating strands of energy can play."[7] A subtle cosmic interconnection between all things in nature, describable mathematically, was an idea envisioned by the great Pythagoras and has been consistently reaffirmed by modern physics. Mathematics nonetheless is not always rational.

THE IRRATIONALITY OF A NUMBER

The proof of the Pythagorean theorem—that in a right-angled triangle the square of the hypotenuse is equal to the sum of the squares of the other two sides—was the epitome of the newly born notion of mathematical deductive reasoning, in which general theorems are proven starting from the least number of axioms. It was especially encouraging to the most important Pythagorean doctrine, "things are numbers." But it ended up also being a bad omen. For soon after the theorem's proof, its application on a special kind of right triangle—the isosceles, with its equal sides having a length of one unit—led to the discovery of a new type of numbers, the *irrational* numbers, which perplexed the Pythagoreans and shook the very foundation of their number doctrine. It was found that the length of the hypotenuse of this right triangle is equal to the square root of two, that is,

$\sqrt{2} = 1.4142135623730950488016887242096980785696718753769480731766797379907324784621070388504\ldots$,

which does not have a precise numerical value; it can only be approximated—shown above it is truncated to 85 decimal places, for there is *literally* not enough paper in the entire universe to write such a number completely! That is, one cannot write down a precise number for the length of such hypotenuse, only an approximate, but we must emphasize that an approximate number is only approximate; it represents not the true length of the hypotenuse but only an approximate length. So how can things be numbers when some things cannot be assigned a precise number? To answer we need to understand irrational numbers a bit more.

In the history of mathematics, integers (. . . , -4, -3, -2, -1, 0, 1, 2, 3, 4, . . .) were supposed to be the only numbers needed, since with their various ratios (fractions) every number that exists (including non-integers) could, so it was thought, be written down. For example, the positive non-integer 1/3 is expressed as a ratio of two integers, obviously 1 and 3; negative non-integer -5/4 is the ratio of the integers -5 and 4; even 0 may be thought of as the ratios 0/2 or 0/7, and so on; in fact even integers themselves may be expressed alternatively as a ratio of two integers, for example, 8 = 16/2. Numbers that can be expressed as ratios of integers are called rational. So, for the Pythagoreans (and in general, up to that point in history) only rational numbers were thought to exist.

Since for the Pythagoreans every number was expressible as a ratio of two integers, so, too, should the length of *every* geometrical line. But they were shocked to discover that the length of the hypotenuse of the aforesaid type of isosceles right triangle could not be expressed as a ratio of two integers! That length was not a rational number. It was equal to the square root of two ($\sqrt{2}$), which turned out to be an irrational number. Irrational *literally* means that there is no ratio, none at all, that can provide an exact numerical value for $\sqrt{2}$. Hence the $\sqrt{2}$ can only be approximated. For example, truncated to one decimal place, the $\sqrt{2}$ is equal to the number 1.4 (which, in this approximate value, can be thought of as the ratios 14/10 or 7/5); to two decimal places, the $\sqrt{2}$ is equal to 1.41 (which, in this approximation, can be thought of as the ratio 141/100). There are infinitely many irrational numbers, all numerically inexpressible by ratios. The famous number π (pi) is irrational.

How, then, can all things be numbers if some things cannot be given an exact numerical value? They cannot if exact numbers is all that we have in mind. But they can in some other broader sense, that of the most advanced theory of matter: the quantum theory. For according to this theory the microscopic particles of matter (quarks and leptons) are regarded as mathematical forms.[8] These forms are really the solutions to the equations of quantum theory and are useful for calculating numerically the average values of various particle properties (e.g., position, velocity, and energy). Since every macroscopic object in nature is composed of these microscopic particles (which are regarded as mathematical forms that generate numbers), then indeed all "things are numbers."

Irrational numbers have been playing a critical role in the advancement of mathematics and physics since the time of Pythagoras. But they still present an epistemological challenge because they provide only a numerically *approximate* knowledge of nature. This "approximate knowledge" is a significant point that will be picked up again in chapter 14, "Zeno and Motion" in order to try to understand the fascinatingly well-reasoned but paradoxical view of Zeno, that apparent motion is not real—that, an apparently flying arrow, for example, is not really moving!

The irrationality of the $\sqrt{2}$ was so shocking to the number doctrine that, according to legend, Pythagoras's student Hippasus of Metapontum (from fifth century BCE), said to have discovered it, was drowned in the deep sea in an act of divine retribution.

Mathematics represents timeless and universal truths, as does nature. And even when the law of a natural phenomenon has not yet been discovered, the law is assumed to exist, as is the mathematical equation that can express it. This in fact is the very premise of science. Without such an attitude science cannot be done and truth cannot be found.

GEOCENTRIC VERSUS HELIOCENTRIC: THE RELATIVE TRUTH

Pythagorean Cosmology

Being a great geometer who understood well the relationships of spheres, flat surfaces, and lines, Pythagoras was probably the first to deduce that the earth is spherical. Several observations might have aided him in reaching such a conclusion. During a lunar eclipse the shadow of the earth on the moon is a circular arc. The masts of receding ships disappear last (and, equivalently, appear first when ships are approaching). Pythagoras himself knew that the evening and morning "stars" are really the planet Venus.

But the most notable achievement of the Pythagoreans in cosmology (often credited to the Pythagorean Philolaus [ca. 470–ca. 385 BCE]) was when they displaced the earth from the center of the universe and imagined it in motion. So, the earth revolves around a center occupied by fire, called Central Fire, and so do the moon, sun, planets (Mercury, Venus, Mars, Jupiter, Saturn—the ones known to antiquity, as only these are visible without a telescope), and the fixed stars—termed so because of their apparently fixed position with respect to one another; on the other hand *planet* means literally "wanderer" because planets were changing their position among the fixed stars.[9] Central Fire is invisible because the inhabitable hemisphere of the earth faces always away from it, whereas the side of earth that always faces it is uninhabitable because it is too hot. Incidentally the moon's synchronous motion—according to which its rotational period around its axis is the same with its revolution period around the earth—produces the same effect: the near side of the moon always faces the earth, whereas its far side always faces away from earth, making it always invisible to an earthbound observer. Revolving around Central Fire is another body, the anti-earth, termed so because of its position.[10] It is imagined to always be in the same direction as the uninhabitable hemisphere of earth, so, like Central Fire, it, too, is invisible. It is not certain why anti-earth was required (some scholars speculate that it was needed to explain eclipses), or even whether anti-earth was really a planet at all—for due to its position anti-earth might have simply been the uninhabitable hemisphere of earth.

In addition to its revolution around Central Fire, earth also rotates around its own axis daily, accounting for the apparent revolution of the sky. This understanding was in audacious opposition to the popular view of an immobile earth at the center of the universe as well as to the evidence of the senses that do not feel earth's motion. In an analogy, to understand the apparent revolution of the sky, pretend to be the earth and stand at the center of a room. Then begin to rotate around the axis of your body, say, counterclockwise. The walls, which you can think of as the sky (with the sun and stars), appear to revolve around you in the reverse direction, clockwise.

Only Central Fire is self-luminous; all other bodies are shining with reflected light from it. In fact this might be the justification of its postulated existence: since the moon is shining with reflected light—an ancient knowledge—and by the cycle of day and night, so, obviously, is the earth. But why not the sun, which in many ways is like the moon—in motion, shape, size, eclipses, color—and all heavenly bodies? If yes, a source of light had to be speculated, hence the Central Fire.

That the Pythagorean system is neither geocentric nor heliocentric is actually a quite justifiable cosmological theory. Since every visible celestial body appears to be moving, perhaps so should the earth, the Pythagoreans might have thought. Now the difficulty with imagining the earth in motion has its origin in our deceptive senses, namely our eyes. Our inability to detect the actual distance of what we see, in particular the stars, is tricking us into thinking that all bodies are the same distance from us. Therefore, being also in different directions, they appear fixed on a hemispherical dome, which is part of what we call the sky. As the sky appears to revolve daily around us the new stars brought into view appear, for the same reason, to also have the same distance from us as all the rest. Thus we imagine every star fixed on a spherical sky—even though at any one time we see only a hemispherical sky—with us on earth at its center. In addition to that, the apparent daily revolution of the dome-like sky around us is easily tricking us into thinking that the earth is absolutely motionless at the sky's center and therefore is at the absolute center of the universe, as if the earth occupies a special position in the universe. This is in fact the geocentric view, which, due to our imperfect senses and our initially uninformed intellect, naturally emerged as the first cosmological model. But the Pythagoreans were well aware of the unreliability of the senses, and they were also accomplished mathematicians with sharp, critical minds. Moreover, being people of virtues, such as humbleness, the Pythagoreans had no difficulty displacing the earth and themselves from the center and purpose of the universe.

No Special Center

Influenced by the Pythagorean cosmology, post-Socratic Greek philosopher Heraclides Ponticus (ca. 390–ca. 310 BCE) proceeded to devise his own. He explained correctly the varying brightness of Mercury and Venus as the result of their varying distances from earth. The additional observations that these planets seem to always follow the sun—when visible, each of them (but independently of the other) either rises just before the sun or sets just after the sun does, especially so Mercury because it is closer to the sun—prompted him to imagine these two planets revolving around the sun—thus justifying their varying distance from earth and consequently their varying brightness—and the sun revolving around an immovable earth at the center of the universe. This partial heliocentric view—with only two planets revolving directly around the sun—became a full-blown heliocentric theory when Aristarchus proposed that all planets including earth revolve around the sun.[11] While this theory was rejected in favor of the geocentric view, it was revived much later by Copernicus.

Traced back to Pythagorean cosmology are the first steps away from the prejudices of the geocentric and anthropocentric worldview and the inspiration for the discovery of the heliocentric worldview. However, perhaps due to scientific misrepresentation of the topic, the popular perception is that the heliocentric model is correct and the geocentric model incorrect. But the profundity of the heliocentric model is really this: (1) it is *another* point of view *as good as the geocentric*—though initially, like the geocentric, it, too, was incorrectly perceived as absolute, as if the sun were the absolute center of the universe (as the *Dialogue Concerning the Two Chief World Systems* of Galileo Galilei [1564–1642] was implying)—and (2) since another center is as good as the previous one, the notion of an absolute center of the universe is abolished. In fact, in modern physics the any-center view is correct. A particular center is chosen merely for its conceptual and mathematical convenience for the understanding of a physical phenomenon and is not to be misinterpreted as absolute or uniquely correct. This view is supported by special relativity (see next subsection). It is also supported by astronomical observations, including the discovery of numerous new galaxies, each with billions of stars revolving relative to the galaxy's center, and generally observations indicating that the universe is isotropic, thus no one location is more special than another. Finally, such view may be accepted based merely on pure humility, that neither the earth nor the sun should occupy a special center, and in general that no point in the universe should be more centered or privileged than another. The universe has neither an edge nor a center, and the laws of physics apply equally the same everywhere. "The merit of the Copernican hypothesis [that (1) annually *earth* revolves around the sun, not the sun around the earth and (2) diurnally *earth* rotates on its axis, not the sky around earth] is not *truth*, but simplicity; in view of the relativity of motion, no question of truth is involved."[12] Equally correct (as will be emphasized a bit more at the end of the next subsection) we could imagine that (1) annually either the earth revolves around the sun or the sun around earth and (2) diurnally either earth rotates on its axis, or the sky revolves around earth. Space and time were absolute in Newtonian physics but became relative in Einstein's theory of special relativity. This means that for relativity an *absolute* frame of reference—a special location of observation that can be used to refer to absolute motion—does not exist. There exist only *relative* frames of reference that can be used to refer to relative motion. Hence we can choose any center *relative* to which something can be at rest or in motion. But a special center for absolute rest or absolute motion is utterly meaningless. Space, time, and motion are all relative. Let us elaborate.

Newton versus Einstein

In Newtonian physics space and time are absolute and thus independent of an observer's relative motion. This means that space distances and time intervals are unchanged by motion. For example, the length and mass of an object are the same for all observers independently of their location or motion relative to the object or relative to one another. The same for them is also the way time passes. Twins, for instance, have always the same age with respect to one another, whether they move or not relative to each other. In general, two events that are simultaneous for one observer are simultaneous for every observer— absolute simultaneity. Space is a kind of preexisting passive (unaffected, in a sense "disconnected" from everything else) playground where objects exist and events occur while time flows steadily in the background the same exact way for everyone (so time, too, is unaffected by everything else). But Einstein's theory of special relativity proved all these to be false, in spite of the fact that all these are how we experience the world daily.

In special relativity the speed of light in a vacuum, designated c, is always 671 million miles per hour—put differently, in one second light travels as far away as is the distance of eight times around the earth. It is the same in all reference frames (for all observers, moving or not relative to the light source). It is also a kind of cosmic speed limit, for although it could be approached, absolutely no material object can travel as fast as or faster than light. This fact has nothing to do with engineering. It is not because we don't have high-powered engines to accelerate an object to the speed of light; rather, this fact is how nature behaves. It is a law of nature that has withstood the scrutiny of experiments since 1905, the year it was postulated by Einstein in his theory of special relativity.

Two of the most dramatic consequences of the constancy of the speed of light concern space and time: they are no longer absolute, they are relative—dependent on an observer's relative motion. They are combined mathematically (by the so-called Lorentz transformation) into a continuum called spacetime. Space distances and time intervals *do* change with respect to an observer's relative motion. Relative space means that a moving object contracts in the direction of motion, as seen by (relative to) a stationary observer—a phenomenon known as length contraction. Relative time means that the passage of time in a moving clock (say, aboard a moving spaceship) is dilated; it is slower relative to the passage of time in a stationary clock on earth—a phenomenon known as time dilation. To function properly the Global Positioning System (GPS) takes time dilation into consideration. If it didn't, the GPS receiver in your car would miss your destination. Parenthetically, the GPS must also take into account another time effect, predicted by general relativity. That clocks in orbit, where gravity is weaker compared to the ground, run faster relative to clocks on earth.

Interestingly, time dilation makes time travel possible because using it we can travel into the future. Suppose Earthly and Heavenly are twins. Heavenly likes to journey in space, while Earthly prefers to stay on earth. If Heavenly travels at a speed close to c upon her return to earth she will realize that she has aged less than Earthly (and all the other people or things on earth). How much less depends on the duration of her trip as well as how close her speed was to c. For example, if her speed was 99.5 percent of the speed of light, then for every one year that Heavenly ages during her trip, Earthly ages ten years. But if her speed was 99.99 percent of the speed of light, then for every one year that Heavenly ages, Earthly ages seventy-one years. So a twin who takes a trip into space will age less with respect to the twin on earth. Here we emphasize that Heavenly feels no different, as regards the passage of time, while traveling. The difference in age is noticed when the twins compare notes, for example, meet again.

In general, if you travel at a speed close to the speed of light, the time elapsed for you will be less compared (relatively) to the time elapsed for those not traveling with you. Hence, because of time dilation, you can then travel into the future of those not taking the trip with you; like the astronauts in the 1968 film *Planet of the Apes*, who aged only eighteen months during their near-light-speed journey and returned to find a post-apocalyptic earth where the elapsed time since they left was 2,006 years. So by taking a trip at high speeds you may return to earth at some future century of your choice. You may enjoy great developments of a more advanced civilization in that future century. The downside, if it is too far into the future, none of your familiar people may be alive to welcome you. Would such a trip be worth taking?

There is yet another fascinating consequence of the constancy of the speed of light. It allows us to see the past. In fact we do it all the time. Looking out in space is looking back in time. And the further out we look, the further into the past we see. This is so because starlight takes time to travel from the distant stars to our eyes. The speed of light is not infinite, so light messages are not transferred instantly. The light from the sun, for example, takes about eight minutes to reach our eyes. This means that observing the sun at, say, 12:00 noon is actually seeing how the sun was at 11:52 a.m. But Polaris, the North Star, is about 434 light-years away from us (i.e., its light takes 434 years to reach us). So looking at Polaris tonight is actually seeing how Polaris looked 434 years ago. Polaris may not even be there now!

There are still other effects of special relativity. A moving object becomes more massive relative to an identical one at rest. Also events that are simultaneous for one observer may not be so for another observer at a different location and/or in motion relative to the first observer. All these so-called relativistic effects become evident only at high speeds though, those comparable to the speed of light. Because the everyday phenomena involve speeds so much smaller than the speed of light, we are tricked into thinking that Newtonian physics is true. It is nevertheless an excellent approximation of truth, for at the limit of low speeds the equations of special relativity reduce to the Newtonian ones. The equivalence of mass and energy (expressed by the equation $E = mc^2$), length contraction, time dilation, and the relativity of simultaneity, are some of the most startling consequences of special relativity.

In light of special relativity, for which space and time are relative, as regards annual motion, according to the geocentric model, relative to the earth (that is, relative to an earthbound observer) the sun appears to revolve around the earth in one year. But equally correct, according to the heliocentric model, relative to the sun (that is, relative to a hypothetical sun-bound observer) it is the earth that appears to revolve around the sun in one year. Likewise, as regards diurnal motion, the sky (with the sun and stars) appears to revolve westward relative to the earth daily (and so the sun appears to rise from the east and set in the west relative to the earth daily)—recall that in our earlier analogy the walls appear to revolve clockwise relative to you. But equally correct, the earth appears to rotate eastwardly on its axis relative to the sky daily—you appear to rotate counterclockwise on your axis relative to the walls. This difference (of what is moving with respect to what) "is purely verbal; it is no more than the difference between 'John is the father of James' and 'James is the son of John.'"[13] The view of the daily eastward rotation of the earth (relative to the sun and all other stars in the sky) is more economical (for only one object, earth, is in relative motion) than the view of the daily westward revolution of the sun and of the myriad stars of the sky (relative to the earth).

CONCLUSION

Within the context of the most advanced theory of matter, the quantum, "things are numbers" indeed. What's more, many apparently unrelated things (phenomena) have already been unified; they are found to obey the same fundamental mathematical equation and thus the same natural law (e.g., the electroweak unification). These findings point clearly to the subtle cosmic interconnection (of mathematical nature) anticipated by the Pythagoreans. But in addition to the aid of mathematics, to find the Logos (reason) of such inconspicuous connections, one needs to be unconventional, to be able to unite diverse fields of knowledge, and to focus a keen eye on the elusive. For only then may one unveil the common characteristics that different phenomena have in all of nature's changes, the perceptible but also the discreet.

CHAPTER 12

HERACLITUS AND CHANGE

INTRODUCTION

Everything is constantly changing, and nothing is ever the same, Heraclitus (ca. 540–ca. 480 BCE) proposed, and in accordance with Logos, the intelligible eternal law of nature. Thus everything is in a state of becoming (in the process of forming into something) instead of being (reaching or already being in an established state beyond which no more change will take place). This means that things, *permanent* things, no longer exist—for they contradict his theory of constant change—only events and processes exist. His doctrine has found strong confirmation in modern physics, for, according to it, absolute restfulness and inactivity are impossibilities; and all happenings, it is speculated, are consistent with a single universal law.

The philosophy of Heraclitus is often stated as aphorisms and has cosmological, ontological, and anthropological significance. The depth of his thoughts as well as the ambiguity of certain concepts he used, such as Logos, fire, and god, earned him the characterization as a dark and enigmatic philosopher. When Socrates was asked to comment about Heraclitus's treatise he replied, "What I understand is excellent; what I don't probably is too, but it would take a Delian [skillful] diver [of the intellect, like the divers from the island of Delos who dive in the deep sea] to recover it."[1]

STRIFE AND HARMONY

For Heraclitus everything in nature is characterized by opposites that are struggling. "We must recognize that war [the competition between opposites] is common, strife is justice, and that all things happen according to strife and necessity."[2] So without strife, as Homer had wished, the universe would be led to its destruction because events and processes could not have existed without some force that promotes change: "Heraclitus criticizes the poet who said, 'would that strife might perish from among gods and men' [Homer *Iliad* 18.107]; for there would not be harmony without high and low notes, not living things without female and male, which are contraries."[3] Hence "strife is justice" because change, for Heraclitus, is caused by the strife of the opposites. Without strife change would not occur.

Now, like Anaximander, Heraclitus, too, requires cosmic justice by such strife. In fact, he argues that not only is absolute dominance not allowed by any of the opposites but quite the reverse, that harmony is born from their strife. "Attunement [harmony] of opposite tensions, like that of the bow and the lyre."[4] This harmony of strife is the result of a subtle underlying unity shared by the opposites; generally, it is the result of the common characteristics that different things have. For example, the property of mass is common to both the different objects the earth and the sun. As a result (according to Newton's third law discussed below) each body attracts the other with the same strength! Discovering and understanding such unity is understanding Logos, but to manage this is difficult because "nature loves to hide."[5] Nonetheless Newton and several scientists thereafter (such as those who created quantum theory) have managed it.

ACTION-REACTION

Newton's action-reaction law, his third law of motion, describes the strife of opposite forces but also their subtle unity and harmony. According to it, for every action force there is an equal reaction force in the opposite direction. For example, the force exerted on a nail by a hammer has the same strength and is in opposite direction to the force exerted on the hammer by the nail. The competing opposites are the competing forces acting in *opposite* directions, but they do so with *equal* strength—so their unity in strength is mathematically expressible, in other words, action force = reaction force. A force (the action) cannot exist by itself; it exists only in relation to its opposite (the reaction)— thus Homer's wish to eliminate strife is unrealizable in Newtonian physics. In fact it is generally shown in physics that "physical action always is *inter*-action, it always *is* mutual."[6]

To appreciate the depth of Newton's third law—in light of Heraclitus's philosophy of the hidden unity of opposites in strife—it suffices to discuss a few more examples: (1) in Newtonian gravity, the earth is attracting you *downwardly* with the *same exact* force as you are attracting the earth *upwardly*! To find out how strong this mutual force is, just jump on a scale and read your weight. Even a freely falling object attracts the earth upwardly *as strongly* as the earth is attracting it downwardly. However, because the mass of the earth is huge, its upward acceleration (toward the object) is negligibly small, but the object's downward acceleration (toward the earth) is noticeable. (2) Analogously, the earth and sun attract each other with forces of *equal strength* (unbelievable but true) and opposite directions, and as a result both celestial bodies move harmoniously through space and time. (3) Both bow and lyre (in the example of Heraclitus) obey Newton's third law, too. In a bow, the cord is pulling each of the two limps (the flexible upper and lower parts of a bow) in one direction—the action force, which is along the cord and toward its midpoint. Whereas the limps respond by pulling the cord in the opposite direction—the reaction force, which is also along the cord but away from its midpoint. Furthermore, action and reaction are forces of equal strength. So the bow's apparent rest is really the result of the constant strife between opposite tensions, the action and reaction. Newton's third law applies even while the cord is being drawn in order to shoot an arrow (or as the cord is being released and shooting the arrow); or as the strings of lyre are at rest, or as they are plucked, producing their sweet notes of music. Even more impressive is that the apparent inactivity, at the macroscopic level, of the bow or lyre at rest (or *any* other object), is, at the microscopic level, really a frantic and endless activity of particle exchange; for force, on the microscopic level, is really an eventful process. And constant change, even the imperceptible, is indeed a fact.

FORCE IN QUANTUM THEORY

According to the standard model of quantum theory the forces of attraction or repulsion between the particles of matter (the quarks and leptons) are caused by the constant exchange of particles of force—called force-carrying particles or messenger particles since they carry the message of the force. The exchange of force particles transfers energy between the particles of matter, causing a change in their own energy, speed, and direction of motion and making them attract or repel.

The massless photons mediate the electromagnetic force; the massless gluons transfer the nuclear strong force (gluons, "glue," bind the quarks to form protons and neutrons for example); the massive W^+, W^-, and Z^0 particles (of positive, negative, and zero electric charge, respectively), the nuclear weak force; and the massless gravitons are speculated to mediate gravity.[7] As regards the gravitons, a theory that describes them has yet to be discovered, and, equally importantly, no experiment so far has confirmed their existence.

The electric repulsive force between two electrons, for instance, is mediated by the continual exchange of photons that, traveling at the speed of light, are emitted and absorbed by the electrons. Namely, one electron rebounds by emitting a messenger photon, and the other electron rebounds by absorbing the photon. Repeated processes of this kind mean that the exchanged photons knock the interacting electrons further and further apart. It is this continual exchange of photons that manifests itself as the electric repulsive force between the two electrons. Similar processes can explain the other forces.

Through the continual exchange of the particles of force, the particles of matter move nonstop and combine with one another to form atomic nuclei, atoms, molecules, and composite objects like bows and lyres. Thus even an apparent static equilibrium of an object at the macroscopic level, down to the microscopic level, is really an eventful, complex, and endless process of particle exchange. Nature *is* constantly changing.

LOGOS

Newton's third law of motion or the more detailed description of a force by the standard model may be viewed as part of Logos. In the third law the underlying unity is the equality of the strength of the opposite forces. In the microscopic interpretation of force, unity is expressed by the conservation laws obeyed by the particles through their interactions (strife); that is, as the particles of matter collide with the particles of force, their net energy (or momentum, to name just two properties that are conserved) before collision *equals* their net energy (or momentum) after collision—again, the Heraclitean unity between competing opposites, is expressible mathematically in physics, that is, energy before = energy after, or, momentum before = momentum after. Of course the actual equations are more descriptive, detailed, and written with mathematical symbols.

Also, matter and antimatter are opposites in strife. Their Logos are the various mathematical laws that they obey, including gravity (the Newtonian concept or Einstein's theory of general relativity), electromagnetism (of Maxwell or quantum theory's), the standard model, string theory, and so on. And the underlying unity consists of the various conservation laws with which each process involving matter and antimatter must comply. The resulting harmony in the strife of matter and antimatter is the general organization of the world (a notion to be revisited at the section "Organization").

In modern physics we are striving to understand various phenomena, first by isolating them and finding which laws they obey. But as in Heraclitean philosophy, according to which true understanding is achieved by identifying common characteristics that different things have, the real picture emerges only when we manage to connect our understanding of isolated and seemingly different phenomena and discover the bigger truth, the Logos they all obey. In modern physics one of the key scientific principles, which is part of Logos, is the Heisenberg uncertainty principle. It will help us understand the doctrine of Heraclitean change from within the context of quantum theory.

THE UNCERTAINTY PRINCIPLE

The most consequential, mind-boggling law of quantum theory—its very heart and soul—is the Heisenberg uncertainty principle. This principle discusses how nature limits our ability to make exact measurements regardless of how smart or patient we are or the sophistication of our experimental apparatus. Namely, as a consequence of the very act of observation, the observer disturbs the object being observed a certain minimum way, causing the result of a measurement to be uncertain. We can measure very accurately the position and velocity of a large-mass object, such as a car or a planet, without significantly disturbing it. We can watch it move and even predict its path of motion. But if instead we had a small-mass object, such as a microscopic particle—an electron, a proton, an atom, even a molecule—we could not measure exactly both its position and its velocity; nor could we observe it in a path of motion or predict its path. Before the uncertainty principle was discovered, absolute accuracy in a measurement, at least in theory, was considered axiomatic, but not anymore.

Suppose we want to observe an electron, hoping to "see" where it is and determine how fast it is moving. To do so we, the observer, must shine light of certain wavelength ("color") upon it—bounce a photon off it. The light (the photon), which is scattered by the electron, will then enter our microscope, be focused, and be seen by our eye. It is the scattered photon that we actually see in an act of observation. Now to illustrate how the observation itself creates the uncertainty in a measurement, we discuss such act in two steps. Step one discusses what happens to the electron when light is shined upon it—when the photon collides with it. Step two discusses how clearly the electron can be seen through the microscope. It is the combination of the effects from these two steps that produces the celebrated uncertainty principle.

Step One: The Collision

As a result of their collision, the bouncing photon transfers some of its energy (and momentum) to the electron and disturbs it (much like when one billiard ball disturbs the motion of another when they collide). But there is no law that can determine the amount of energy imparted on the electron by the photon.

Thus the photon pushes and disturbs and changes the velocity of the electron unpredictably. This means that the electron may have a range of possible recoil velocities, hence its velocity cannot be known precisely: there is an uncertainty in its velocity. On the other hand, the disturbance introduced by a photon bounced off a car or a planet is undetectably small because, compared to an electron, the mass of a car or a planet is huge; just think how much more difficult it is for anyone to push and disturb a real car, which weighs a lot, compared to pushing a toy car, which does not weigh much. The velocity and location of a car or planet can be measured almost with absolute precision. This is in fact another reason that classical physics (e.g., Newtonian physics), which does not include the uncertainty principle, works quite well for macroscopic objects.

Now, concerning the electron, we can reduce the uncertainty in its velocity by using a photon of smaller energy so that its push to the electron is gentler. But a photon's energy is inversely proportional to its wavelength: the smaller the energy, the longer the wavelength (the "redder" the color is), a relationship that brings me to step two. Unfortunately, while a photon with a longer wavelength has less energy, which reduces the uncertainty in the velocity of the electron, it simultaneously increases the uncertainty in the position of the electron—the image of the electron gets fuzzier. Why?

Step Two: The Microscope

Because the determination of the position of the electron depends on the wavelength, too. This dependence, which is known as the resolving power of a microscope, regulates how clearly something can be seen—how well the scattered light can be focused and thus how accurately the electron can be located. The longer the employed wavelength, the fuzzier the image of the electron will be, and the greater the uncertainty in its position. What we see through the microscope is really a fuzzy flash created from the photon scattered by the electron. The electron, which is a point-particle, is somewhere within this flash, but where exactly is indeterminable. Its position cannot be known precisely. The flash may be focused into a region no smaller than the wavelength of light (the law of the resolving power states). Hence the uncertainty in the position of the electron may be equal to or greater than the wavelength of light, but never smaller than it! So at best, the minimum uncertainty is equal to the wavelength. Since the position cannot be known precisely, the electron has a range of possible locations it can occupy, just as it has a range of possible recoil velocities to move with.

Given that light of zero wavelength does not exist—that is, we cannot observe if the light source is turned off—the uncertainty in the position can never be zero—we cannot see with *absolute* precision where the electron is. Nonetheless, we can reduce the uncertainty in the position by using light of a smaller wavelength, though unfortunately this action simultaneously increases the uncertainty in the velocity—for as seen in step one, the smaller the wavelength, the greater both the light energy and the disturbance imparted on the electron (i.e., the greater the range of possible recoil velocities).

Position-Velocity Uncertainty

The wavelength of light used in an observation has conflicting effects; there is a trade-off in the determination of the position and velocity of a particle. The result is the position-velocity uncertainty principle: the more precise the position, the more uncertain the velocity, and vice versa. Heisenberg proved that the product of the two uncertainties can never be less than a certain minimum positive number—which is roughly equal to Planck's constant, a fundamental constant of nature, divided by the mass of the particle.[8] Consequently, *absolutely* precise knowledge of either property is unattainable because if one of the uncertainties were zero, their product would also be zero, a result that would be in clear violation of the principle. In classical physics, on the other hand, of which the uncertainty principle is not part, these uncertainties could each be zero—thus a particle's position and velocity could, at least in principle, be determined exactly—leading to what is known as classical determinism, which is the opposite of quantum indeterminism (that is, quantum probability), the consequence of the uncertainty principle.

CLASSICAL DETERMINISM VERSUS QUANTUM PROBABILITY

In the macroscopic world of classical physics, by knowing the forces that act on an object as well as the object's exact position and velocity at some initial time, we can determine its exact position and velocity (its trajectory) for all future time. So its motion is precisely determinable: a path can be plotted, even watched live, point by point continuously from an initial instant to any future one. Because of this capability classical physics is said to be deterministic. We can plot the precise orbit of a space shuttle, for example, just by knowing the forces acting on it and the initial conditions (its position and velocity at some initial instant), and we can watch it fly through space and time as predicted by our equations. It is therefore easy to predict a solar eclipse—when the earth, the new moon, and the sun will align—but absolutely impossible to measure or predict where an electron in an atom is or will be. Why?

Because, according to quantum theory, the subatomic world of particles is profoundly different than everyday experience; it cannot be described by classical physics. Inherent in the uncertainty principle, which limits the accuracy of a measurement, particle properties (such as position, velocity, momentum, and energy) cannot be assigned an exact value, neither initially nor at any time later. Thus they must unavoidably be expressed only as probabilities, which then lead to quantum indeterminism. Solutions of the so called Schrödinger equation can be used to calculate such quantum probabilities—for example, the probability of finding a particle at a certain location at a certain instant of time. A probability is a number that represents the tendency of an event to take place, not its actual occurrence. Hence, the best we can do is theoretically to predict only *probable* outcomes, and experimentally measure *also* only probable outcomes. Consequently, a particle's path of motion can neither be predicted (nor plotted), nor can it be observed; it is indeterminable: establishing a definite, traceable, point-by-point orbit is an impossibility. In fact, the very notion of an orbit is inadmissible in quantum theory. The determinism of classical physics is therefore replaced by the probability of quantum theory. And the consequences of this fact are staggering. The true nature of nature, for example, is different than the way it appears to be through mere observation.

OBSERVATIONS ARE DISCONNECTED EVENTS

During the act of observation all we see through the microscope is a flash of light somewhere within which the particle exists. But where exactly it is within this flash at each instant, and what it does when we are observing it, whether at rest or in motion, are all indeterminable. Even worse, nature does not allow us to know what happens between consecutive observations. Consecutive observations have time and space gaps; flashes are seen one at a time and spatially separated. Hence, inherent in the uncertainty principle observations, *any* observations of both the microscopic and macroscopic world, are always disconnected events! Roughly speaking, it is as if we are observing nature by continuously blinking.

This is a profound result and in direct contradiction with apparent reality according to which the changes in the daily phenomena are observed to occur continuously. The act of watching an arrow in flight, for example (an interesting thought experiment to be revisited in the section "The Arrow Paradox" in chapter 14) is really a series of disconnected observations, which to our imperfect eyes appear to occur continuously only because the time and space gaps between subsequent observations are undetectably short. The arrow's apparent continuity of motion is therefore an illusion. The shortness in these gaps is, incidentally, a consequence of the fact that in an observation, the disturbance introduced by a photon bounced off a macroscopic object such as car or an arrow is undetectably small because these objects have comparatively more mass than the mass of microscopic objects such as electrons and protons. This is, recall, also the reason that macroscopic objects have (actually, appear to have) definite orbits while microscopic objects do not.

So observing anything, anything at all, can happen only discontinuously. It is, roughly speaking, like cinematography (motion pictures), where a series of separate drawings, each, say, of a ball at a different position, is flashed before us rapidly (with short time gaps). Now, (1) if the position of the ball is changed *gradually* in each subsequent drawing, that is, the distances (the space gaps) between each new position of the ball and the previous one are sufficiently short, then, when the drawings are flashed before us, the ball is observed to move continuously (thus with a definite orbit). But this continuity in observation is really an illusion of the deceptive senses that cannot notice the short gaps. Case (1) corresponds more to how we observe macroscopic objects. On the other hand, (2) if the space gaps are sufficiently long, then, when the drawings are flashed before us, the ball is observed to move discontinuously. Case (2) will correspond more to how we observe microscopic particles, but only after the following two modifications: first, do not think of the ball to be the actual particle; rather, it roughly corresponds to the flash of light somewhere within which the observed particle exists; and second, as we will argue in "Nature as a Process" below, even if we do observe a similar *type* particle (say, an electron) at consecutive observations, it is indeterminable if it is the same *particle* (electron), even when the time and space gaps between observations are short. These two modifications, which capture more accurately how we observe microscopic particles, make it impossible to plot a definite path of motion for any microscopic particle.

Now, the reason that the phenomena are observed to occur discontinuously might be that the very phenomena *themselves* occur discontinuously (even when we are not observing); they may not just be *observed* to occur discontinuously. In any case, the discontinuity in observations has astounding consequences: in the section titled "Nature as a Process" we will use it to question the very identity of a particle, and in chapter 14, "Zeno and Motion," we will use it to question the reality of motion *itself*.

CHANGE

The Heraclitean doctrine that everything is constantly changing and nothing is ever the same has three implications. First, there is a change; second, the change is constant; and third, because nothing is ever the same the constant change is unidirectional. Modern physics agrees with all three: first, change occurs in two different ways: (1) through motion and (2) through the transformations of matter and energy; second, the uncertainty principle of quantum theory as well as general relativity affirm that change is constant, and in addition, as seen just above, quantum theory ascertains it is also discontinuous; and third, the second law of thermodynamics discusses how change is unidirectional—the universe becomes increasingly disordered.

(1) Motion Causes Change

Change caused by motion is discussed in the following three cases.

A. The motion of the particles of matter causes their rearrangement in an object and consequently causes change in its various qualities (e.g., density and temperature). For example, atoms are more compressed in denser objects and jiggle faster in hotter objects.

B. Space itself is changing because matter distorts it, a phenomenon that can be understood when we describe how gravity works within the context of the theory of general relativity.

Gravity, in general relativity, is explained by giving space properties, namely, by regarding it as a flexible medium distorted by matter—like a trampoline surface that is stretched and warped by a bowling ball resting or moving on it. In the case of the earth and the sun, for example, the distortion of space caused by one body influences the motion and is felt as gravity by the other.

In a simplified analogy, the flexible trampoline fabric (which plays the role of space) is curved when a bowling ball (which plays the role of the sun) rests on it. The geometry (shape) of the fabric depends on (1) the mass of the bowling ball and (2) the distance from it: (1) the more the mass, the more curved the fabric (space) becomes; (2) the closer to the bowling ball, the greater is the curvature of the fabric. The distorted fabric in turn influences the motion of a small marble (which plays the role of the earth) rolling on it. Depending on how we start the marble moving (i.e., with what initial speed and direction, and from what location), it will move on the distorted fabric around the bowling ball by following a particular path (circle, ellipse, parabola, spiral, straight toward the bowling ball, etc.), and thus will appear to be attracted by the bowling ball. For example, a marble released from rest moves on the distorted fabric caused by the bowling ball and plunges onto it, like an apple falls from its tree onto the ground by moving through the distorted space caused by the earth. In the trampoline analogy the distorted fabric *is* gravity; and the greater the bowling ball mass, or the smaller the distance from it, the stronger gravity (the distorted spacetime) becomes.

In the earth-sun case the sun distorts the spatial fabric around it (and time, too—it passes more slowly as one gets closer to the sun). Traveling at the speed of light, this distortion reaches and affects the motion of the earth—analogously, a water disturbance, a water wave, travels in the sea but with a much smaller speed. In turn the earth traverses the distorted space as if space pushes the earth through it. Of course the earth distorts the space around itself, too (although its smaller mass produces a much smaller distortion than that of the sun); and so do the moon, planets, stars, and galaxies. Gravity is really the twists, curves, ripples, bumps, depressions, and in general all these distortions (the changing geometry) of spacetime. And each body's motion is actually a response to the space distortions from all other bodies around it. Because these bodies are in constant motion in the universe, the pattern (the geometry) of space distortions that they create is in a state of constant flux—and so the motion of matter causes change in the geometry of space. In our analogy, as it rolls, the bowling ball transfers the warping of the trampoline surface to different locations. Because both space and time are distorted by matter, spacetime in general relativity becomes a four-dimensional malleable (distortable) continuum. In turn, these spacetime distortions, which we usually call gravity, influence the motion of matter.

C. In addition to its constant warping, space as a whole is also expanding, thereby carrying all the galaxies with it and causing them to move away from each other. Here again, motion, which in this case is a consequence of the expansion of space, produces change. Known as the expansion of the universe this was first predicted theoretically by the solutions of the equations of general relativity, shortly after their publication in 1916. It was later confirmed experimentally by astronomer Edwin Hubble (1889–1953) in 1929 when he observed a redshift in the light emitted by the distant galaxies. The redshift is a measure of the relative velocity between a galaxy and the earth. Specifically, it means that distant galaxies are rapidly receding from us. The greater the distance, Hubble discovered, the faster the recession speed, a result known as Hubble's law. This law is included in the big bang model.

Galaxies are not moving out into preexisting space, a common misinterpretation of the phenomenon of expansion, but they are moving away relative to each other, and what carries them is space itself as it is expanding (stretching); the result is that the size of the universe is increasing with time. Furthermore, the recession of galaxies does not make the earth the center of the universe or in any way a more special place than any other. Quite the opposite, because the universe is isotropic the expansion would look the same from any location in the universe. In a classic analogy, imagine how any dot (i.e., galaxy) on an inflating balloon is seen receding from the perspective of any other dot as the expanding membrane (i.e., space) itself carries all dots with it. Since galaxies are observed as receding from each other and the universe as expanding, in the past they must have been closer to each other, and the universe must have been much smaller—imagine our balloon to be deflating. In the extreme case, the whole universe (all the galaxies, all matter, energy, and space, even time) is imagined to have been a mere point, its explosion of which is the premise of the big bang model. Whether such a point-like universe actually existed is still only a hypothesis, and why it exploded is still puzzling. Nonetheless, we do know that the universe must have been very, very small (if not point-like) when it exploded, as described by the big bang theory, and that it has been constantly expanding (stretching), thus changing ever since.

(2) Transformations Cause Change

Transformations of matter and energy also cause change in the various qualities of objects (or the universe in general) via two types of processes: first, when the various particles of energy convert into particles of matter, back and forth, materializing and dematerializing; and second, when one type of material particle converts into another. An example of the former is the materialization of the energy of light (or invisible gamma rays) into an electron-positron pair, or the dematerialization of such pair into energy (light or gamma rays); and an example of the latter is the conversion of two protons into a proton-neutron nucleus (a heavy form of hydrogen called deuterium), a positron, and an elusive neutrino (all common reactions in the stars energizing them with their light). Of course more everyday-type transformations of matter and energy, such as from solid to a liquid (as the melting of ice) or liquid to a gas (as the evaporation of water) cause change, too.

But things are not merely changing, they are *constantly* changing, a conclusion required by the uncertainty principle.

Change Is Constant

To avoid violating the uncertainty principle, motion in nature must be perpetual. If a particle could sit still, it would mean that its velocity would be exactly zero and so then would be the uncertainty in its velocity. Consequently, the product of the position and velocity uncertainties would also be zero, a result in violation of the Heisenberg uncertainty principle. The principle holds only if motion is perpetual. A particle cannot sit still, ever. This result is also supported by the third law of thermodynamics, which states that the absolute zero—the lowest possible temperature for which every particle in a substance would be motionless—is unattainable (temperature is a measure of particles' average energy of motion). Now, since the motion of particles is perpetual, so is change. Incidentally, since motion is constant, then motion is also involved even when change is caused by the transformations between the particles of matter and the particles of energy.

Support of the "constant change" view comes also from the expansion of the universe that has been happening ever since the big bang. What's more, the spacetime continuum fluctuates constantly at microscopic scales, like a turbulent sea—the distortions of space are changing violently—a phenomenon resulting from efforts to reconcile the theory of general relativity with the uncertainty principle of quantum theory.

Change Is Unidirectional

But change is not merely constant; it is also unidirectional, meaning nothing is ever the same; "you could not step twice into the same river."[9] Heraclitus parallels the constant unidirectional change in nature to the ever-changing waters in a river. If the state of a river at one moment were ever the same as the state at another moment, it would have been possible for one to step twice into the same river. Since one cannot do that, nothing ever is the same, and so change is not only constant; it is also unidirectional. In fact, one "could not step twice into the same river" not only because a river's waters are ever-changing, but also because one's own body is also ever-changing. *Everything* is constantly changing, and *nothing* is ever the same.

Now, according to the second law of thermodynamics, net entropy—the degree of disorder (randomness) in the universe—is always increasing. Thus, nature is in a state of becoming, but it is a disorderly becoming.

Indeed, then, everything is constantly changing, and nothing is ever the same! Heraclitus's doctrine of change includes everything, even the seemingly unchanging, such as a rock in its apparent state of rest or even a human body. In fact, even the gradual biological evolution by descent and variation—that the more complex life-forms do not arise spontaneously but evolve from simpler ones through modifications—is a principle to be expected as a consequence of the Heraclitean theory of constant change.

NATURE AS A PROCESS

Heraclitean View

A profound consequence of the Heraclitean theory of universal constant change is the view of nature as a process made up of events. For the notion of "a thing" is inconsistent with a theory of constant change. To be able to be spoken of and defined, the thing must remain absolutely the same for at least a period of time; it must have some permanence and must be identifiable. But the notions of sameness and changelessness are contradictory to a theory of *constant* change. Consequently, it is more appropriate to consider a thing as an event (something happening somewhere at some instant of time) and not as something permanent. Thus, what changes is not something material or initially permanent; what changes are the events. Groupings of events constitute processes, which in turn make nature the ultimate process.

Quantum View

This notion is supported by quantum theory. We will argue that microscopic particles are better understood to be events rather than permanent things.

We learned earlier that because of the uncertainty principle observations are disconnected events. Now, without continuity in observation, without the ability to keep a particle under continuous observation (even for the smallest time duration), how can we establish its identity or permanence? With time (and space) gaps between observations during (and within) which we cannot see what a particle is doing, how can we be sure whether, say, an electron observed at location A has moved there from location B, or whether it is really one and the same electron as that observed at location B, regardless of their proximity? We cannot! Since observations are disconnected events, consecutive observations of identical particles—such as electrons, all of which have the same intrinsic properties, for example, charge, spin, rest mass—might in fact be observations of two different particles belonging in the same family (e.g., two different electrons), and not observations of one and the same particle (e.g., the same electron) that has endured for a certain period of time. It is therefore impossible to ever determine whether the observations of two identical particles could actually be observations of one and the same particle, and consequently whether a particle endures for a period of time.

So without the ability to keep a particle under continuous observation, it is impossible to establish experimentally its identity or permanence. Because of this, the notion that a particle is an identifiable individual and a permanent *thing* breaks down (or, it is an ambiguous notion, to say the least). The alternative is to consider a particle to be an event.

General Relativity View

Particles, in the view of general relativity, can endure up until they convert to energy. Until then they are identifiable permanent entities because general relativity has not yet been reconciled with the uncertainty principle. Still, particles are events in general relativity, too. This is so because matter is intricately connected with the fabric of spacetime (they are continuously affecting each other). So as time is constantly changing, so are, in general, the properties of space and matter. Hence a point in the continuum of spacetime is regarded as an event. And so a particle occupying a space location at an instant of time is treated also as an event. Two events are separated by their spacetime interval, which involves a spatial distance and a time interval.

In conclusion, matter, energy, space, and time are all intimately linked, interacting with one another constantly, causing changing events and processes. Nature is the perfect process and therefore is in a state of becoming; nothing ever *is*. But what causes change in the theory of Heraclitus, and what causes change in modern physics?

FIRE AND ENERGY

A permanent primary substance of matter is contradictory to a theory of constant change. The only element of permanence in such theory is change itself. What really causes change then? For Heraclitus it was the "everlasting fire,"[10] and for modern physics the eternal energy. The strife of the opposites or the interactions of matter are fueled by fire and energy, respectively. Matter, the events as has been argued, is the result of the transformations of the fire or of the energy. Fire and energy also represent particular processes. They cause cooling and condensing or heating and rarefying or forming and dissolving. "The transformations of fire [energy] are, first of all sea [liquids]; and half of the sea is earth [solids], half whirlwind [gases]."[11] And the transformations of fire, as it is with energy, occur with measure—by obeying conservation laws—since in a metaphor Heraclitus argued, "All things [matter] are an exchange [is a transformation] for fire [of energy] and fire [and energy is a transformation] for all things [of matter], as goods [as one type of matter] for gold [transforms into another type of matter or energy] and gold for goods [back and forth]."[12] In another statement he writes, "This world, which is the same for all, none of the gods nor of the humans has created, but it was ever and is and will be, an everlasting fire [energy], flaring up with measure [conservation] and going out with measure [transforming back and forth between its various forms by obeying the law of conservation of energy]."[13] Change, which in each respective theory is caused by fire or energy, is guaranteed to always occur since fire and energy are conserved while they are also changeable. From all these qualities fire and energy seem to qualify as the primary substance of the universe in each theory. As argued by Karl Popper, if the world is our home, then for Heraclitus fire is not *in* the house, "the house [the *house*] is on fire," a "somewhat more urgent message."[14] Equivalently, we can argue, the world is energy.

Fascinated by the similarities between the Heraclitean fire and energy Heisenberg wrote: "Modern physics is in some way extremely near to the doctrines of Heraclitus. If we replace the word 'fire' by the word 'energy' we can almost repeat his statements word for word from our modern point of view."[15] And also, "Energy is in fact that which moves; it may be called the primary cause of all change, and energy can be transformed into matter or heat or light. The strife between opposites in the philosophy of Heraclitus can be found in the strife between two different forms of energy."[16]

ORGANIZATION

But while everything is constantly changing, and defining a permanent fundamental particle of matter is impossible, something must endure, at least for some time. For, in all this constant change, still, definable "things" such as rivers exist and so do we, and rivers are distinct from us, and one river (or one person) distinct from another. So both we and the rivers are constantly changing, but simultaneously both we and the rivers are recognizable. "We step and do not step into the same river; we are and are not,"[17] said Heraclitus. How can this be?

In a constantly changing nature, which is best regarded in terms of events, each event is different. But *collections* of events can endure, by creating a certain macroscopic average, which, for a period of time, is a recognizable "organization";[18] the identifiable plethora that we call things may then be viewed as different organizations.

From the above Heraclitean quote, in the part "we step . . . into the same river; we are . . ." Heraclitus treats the "we" and "river" as two identifiable organizations, thus as two different and permanent things—at least for some period of time. Whereas in the part ". . . and do not step . . . and are not," he seems to imply that while the "we" and the "river" appear as permanent things these are so only on the average. In reality neither "we" nor the "river" are ever the same.

Heraclitus realized that while a thing (an organization) can on average be identifiable for some period of time, strictly speaking the thing is uniquely different at each instant of time. And so while there may be a collection of events that can produce an identifiable organization called river, which on average persists unchanged for a certain time duration, this river is never ever really exactly the same. Logos (the cause of change, the law of nature), in the view of Heraclitus, is the only thing truly eternal. Although through a truly strict interpretation of a continuously changing nature without anything ever the same, even Logos should be changing. The modern physics equivalent of this is the hypothesis that the fundamental constants of nature, numbers that describe the various laws we have and are the reason the universe is what it is (such as the speed of light, Planck's constant, or the gravitational constant), might after all be functions of time. If this is discovered to be true, then the order and organization of tomorrow's nature (especially in the long run) will be so unknowingly different from today's, a real intellectual treat for those inquiring minds who like the constant search and discovery, for in such a case, the mysteries concerning the nature of nature will be forever changing, and so will our very knowledge about them.

CONCLUSION

Heraclitus declares the being (that which exists, nature) but identifies it with becoming. All follows from that: everything is constantly changing, material sameness is impossible, there is a plethora of different events that make nature a process, and described by warring opposites that nonetheless obey Logos. But Parmenides declares just the *Being*; only what is, is, what is not, is not. All follows from that: change, he argues, is logically impossible and so what is, is one and unchangeable! This dazzling and absolute monism is in daring disagreement with sense perception.

The Heraclitean and Parmenidean worldviews are therefore antinomies (contradictions), for starting from a being the two philosophers developed a unique series of logical arguments and arrived at opposite results: for the Heraclitean, being is becoming, but for the Parmenidean, Being just is. It is Heraclitean change and plurality versus Parmenidean constancy and oneness. But it is a controversial oneness, for Being's exact nature is uncertain.

CHAPTER 13

PARMENIDES AND ONENESS

INTRODUCTION

Philosophy was seriously shocked by the logic of Parmenides (ca. 515–ca. 445 BCE). Being the first philosopher of ontology, Parmenides questioned the nature of existence itself and created his monistic philosophy by contemplating the most fundamental of questions: How can something exist? And, what are the properties of that which does exist? And through purely rational arguments he marvelously reasoned out an answer that overturned completely the common perception of the world around us! In particular he asked, How could there be something instead of nothing? What does it mean to say that something exists? Can existence (nature) come to be from nothingness? Is there such a thing as nothingness? Has nature been caused by a primary cause—that is, by an absolutely first cause that permits no cause (no explanation) of its own? Does nature have an ultimate purpose that permits no purpose of its own? What is the nature of nature? Remnants from his profoundly abstract thought are present in modern cosmological models describing one indivisible and whole universe, unborn, eternal, and imperishable.

IRONCLAD LOGIC

First he argued that we can think only about that which exists, the *Being*, "for the same is the thinking and the Being."[1] On the contrary, he thought, we can neither speak about nor think about something that does not exist, *Not-Being*. For if we could it would mean that Not-Being had properties (those mentioned speaking or thinking about it). But true nothingness is property-less. Therefore, the notion of nothingness (Not-Being) is impossible! This is in fact the

169

critical premise of Parmenides's theory. And to understand his arguments we must always remember that for him what does not exist, does not exist, neither now nor before or after, neither here nor there; that is, we cannot assume what does not exist now (or here), could exist later (or there), or could have existed before somewhere. No! Only what is, is—only Being exists. What is not, is not—Not-Being does not exist.

With this premise in mind he proceeded to figure out if change is logically possible. Change, he maintained, requires that the notion of nothingness exists. But since such a notion is an impossibility, so then is change; Being is unchangeable—for if it could change, it would change into something that Being is not already, into something new that does not yet exist, thus into Not-Being, but this is an impossibility, for Not-Being does not exist ever anywhere. Analogously, if it could change, it would cease to be what it once was, thus what once existed (Being) would no longer exist; it would become Not-Being, but this is again an impossibility for Not-Being does not exist. In other words, that which exists (Being) cannot change because change requires that the notion of nothingness (Not-Being) exist. Because only then could Being have been it (Not-Being) and could have again become it. Equivalently, we can say that change is impossible because it requires that something is either created from nothing or destroyed into nothing, but since the notion of nothingness does not exist, change does not exist either.

Being is also unborn—it is uncaused, that is, it has not been caused by anything, thus has no beginning—and it is imperishable—it has no end, no ultimate purpose. It just is! It could not be born, Parmenides thought, for if it could, it would be born from either (a) Not-Being—but this is an impossibility, for Not-Being does not exist—or (b) Being—also an impossibility, for something cannot be born if it already exists, that is, something cannot be born from itself. Analogously, Being cannot perish; for nothingness, which Being must become in order to perish, does not exist. Hence, what is (Being) just is; it neither comes to be from nothing nor perishes into nothing. This remarkable Parmenidean thesis was embraced by the pluralists Empedocles, Anaxagoras, and Democritus, and as we will see, it was applied in their own theories. Well, then, since Being is unborn, unchangeable, and imperishable, there is neither becoming nor passing away—nature just is!

Being is also always everywhere: there is neither a place nor a time where or when what is, is not already complete (e.g., of the same amount, appearance, and generally of the same properties). For if somewhere or sometime Being were less than complete, if it lacked something, it would mean that somewhere or sometime that something that Being would lack would not exist; it would be Not-Being, but since Not-Being does not exist, there is never any expectation for Being to be it. So Being is always complete everywhere. Hence, diversity and plurality are illusions of the senses. It is also motionless—for being always everywhere, there is never anywhere to go where it is not already. Similar-type arguments lead to the various properties of Being.

The nature of nature (of Being) is of the purest oneness: there is only one thing, Being. It is an indivisible eternal whole, unborn, unchangeable, imperishable, continuous, indestructible, finite, and uniform (always the same everywhere). This is a dazzling but provocative oneness, for it is (or so it seems, anyway) logically sound, yet it is also daringly in stark contradiction to apparent reality. And what its exact meaning really is depends on how these properties of Being are regarded, literally or metaphorically, of material or immaterial nature. Being is that which *is*. All other characteristics beyond that are quite uncertain and debatable, for what *is* (what exists) is *really* the question for Parmenides. But irrespective of its nature, Being captures a highly valued place in pre-Socratic philosophy and modern physics alike—namely, oneness!

THE NATURE OF BEING

Modern physics embraces a kind of monism and wholeness, too, for it tries to ultimately unify all four forces (gravity, electromagnetism, the nuclear strong, and the nuclear weak) and all particles under a single overarching principle in which there will be only one unified force or, equivalently, one type of fundamental particle, suggesting a subtle interconnection and oneness in all apparent plurality. Hence, should the properties of Being be interpreted metaphorically, Being might be a metaphor of the one, unchangeable, universal, eternal, objective truth of nature (a unified force of a theory of everything), which can be discovered through a combination of observation and rational contemplation. If so, Parmenides's philosophy is not necessarily against the notion of change but rather is in support of the view that the true way things change is not at all as perceived by the senses. I believe this view is reasonable since in his philosophy "the same is the thinking and the Being," that is, we can think only about something that exists; then since (1) we think about the observed changes, they must exist, though they must occur much more complexly than they appear (and in fact they do) because (2) we *also think about the complex and subtle ways that these changes occur* (through our scientific models, e.g., general relativity, quantum theory, biological evolution, etc.).

On the other hand, should the properties of Being be interpreted literally, then nature is one, uninterrupted, indestructible, indivisible, eternal, and material whole; it's a kind of full and solid block of matter without parts (uniform). Such a type of full nature implies that there is no void (empty space). The void is Not-Being for Parmenides, true nothingness—it does not exist. But interestingly without the void it is difficult, if not impossible, to accept that motion and change are real. For the easiest way to understand the occurrence of motion and thus change, too, is to imagine the existence of empty space within which things could move. Assuming there is no empty space, motion is an illusion and so is change. The atomists Leucippus and Democritus found this conclusion utterly absurd but inspiring as well. For to create their atomic theory and explain motion and change rationally they had to employ both: Parmenides's Being—its material nature in particular—and his Not-Being.

ATOM AND BEING, VOID AND NOT-BEING

So by giving Parmenides's ideas a straightforward, literal, and material meaning, the antithetical notions of his Being (existence) and Not-Being (nonexistence) evolved in the minds of the atomists into the antithetical notions of "the full"[2] (the atom), and "the empty"[3] (the void, empty space), respectively, and became the essence of their atomic theory. Incidentally, the intellectual continuity in our efforts to know nature is unquestionable in this case. Now, there were many atoms (Beings) with key properties of Being (i.e., whole, indivisible, indestructible, solid, with no parts) and lots of empty space (Not-Beings) within which atoms can move. Interestingly, although the atomists could not counter Parmenides's arguments against the existence of Not-Being, they still identified it with a certain kind of nothingness that *existed*, the empty space. But empty space's perception has been controversial ever since its conception.

NOTHING COMES FROM NOTHING

For Parmenides, there is no empty space, for empty space is nothingness, Not-Being, and Not-Being does not exist. How can something, which is nothing, really exist? Parmenides thought. How can something be defined and assigned properties when it is supposed to be property-less? It cannot, he argued. We are unable to even think of nothingness, he reasoned. Nothingness is a meaningless concept, for if nothingness existed, it would not really be nothingness; it would be something-ness. If we could refer to something and give it properties, that something could not be nothing; it would be something real and would exist.

Parmenides wanted to understand change, motion, and the empty space via purely logical arguments. For the empty space especially, he thought there was no good logical argument in support of its existence. Whether the empty space is a true nothing or not is a notion to be revisited in chapter 17, "Democritus and Atoms." Nonetheless, Parmenides thought that empty space was a true nothing, and as seen above, he also argued that *nothing* comes from nothing—Being neither comes to be from nothing (it is unborn) nor passes away into nothing (it is imperishable).

This principle has in fact a counterpart in modern physics in the notion of energy, which includes mass, since for relativity mass and energy are basically the same, as implied by $E = mc^2$. Like Being, energy can neither be created—it does not come to be from nothing—nor destroyed—it does not pass away into nothing. It just is and its total amount is unchangeable and enduring. Particles of matter and antimatter are constantly being created and annihilated, for example, but not out of nothing and into nothing. To occur these processes require something to already exist—namely, energy. They are created from energy and annihilated into energy. There is no mechanism in modern physics that violates the basic Parmenidean idea that something can neither be created from nothing nor pass away into nothing. All interactions require something, energy (or matter, since they are equivalent), but also space and time. Indeed, nonexistence is an impossibility even in modern physics, and the uncertainty principles of quantum theory (see next paragraph) may be considered as statements in support of that. Why use these principles? Because these principles are relationships between space, time, matter, and energy, concepts that constitute the essence of nature (of something-ness). And if we hope to prove that the notion of nothingness is an impossibility—that nothingness is not derivable from something-ness—well, we had better begin from an analysis of principles that describe the essence of something-ness.

So to argue for this (that nonexistence/nothingness is an impossibility) we first recall the position-velocity uncertainty principle: the product of the uncertainties in the position and in the velocity of a particle must be greater than Planck's constant divided by the particle mass—that is, such product is greater than zero. Analogously there exists the time-energy uncertainty principle. It states that the product of the fluctuations in the energy of a particle and the time interval that the particle endures must be greater than Planck's constant—again such product is greater than zero. For these uncertainty principles to hold spatial distances, time intervals, velocities, and energies are forbidden from ever being absolutely zero—that is, their nonexistence is forbidden. For example, the smaller the confining space of a particle is (or the briefer the time interval the particle endures in such confinement), the more frantic its motion and energy are. But neither the confining space nor the time interval can ever be exactly zero, for if they were, the uncertainty in position and the uncertainty in time would have been zero, too, and consequently both uncertainty principles would have been in violation—the product of the uncertainties in position and velocity, and in time and energy, would have been zero, too (instead of greater than zero). Similarly, both uncertainty principles would have been in violation if a particle had zero velocity or energy. The principles hold only if spatial distances, time intervals, velocities, and energies are nonzero; they must exist; they cannot be nothing! (In a sense such result is expected because our physics relationships, equations or inequalities, are in the first place conceived to describe *something*, not nothing; the notion of nothingness is indescribable.) Hence, as per Parmenides's reasoning and as per the uncertainty principles, nothingness is not only not allowed to exist—for *nothing* comes from it (e.g., the uncertainty relationships are violated and thus cannot be used to account for what exists)—but, equally profoundly, existence is required, that is, spatial distances, time intervals, velocities, and energies must be nonzero (for only then do the uncertainty relationships, which describe something-ness, hold).

In fact, one of the fundamental tenets of quantum theory is that information cannot be lost from the universe (recall the section "Black Holes: Challenges in the Quest for Sameness" in chapter 8). In Parmenidean terms this means that, what is—Being, information already present—cannot become Not-Being—information cannot be lost. So Stephen Hawking might be onto something with regard to his new analysis of black holes.

Quantum theory (the essence of which is the uncertainty principles) is then in accordance with Parmenidean philosophy, "for the same is the thinking and the Being": for we can think only about that which exists, in other words, the uncertainty principles describe only something-ness and forbid nothingness. With this in mind Parmenides's main question, How can something exist? may now be answered: within the context of modern physics, something (Being) must exist because nonexistence (Not-Being) is impossible. Now, what is the nature of that which exists?

AN INDIVISIBLE WHOLE

Relativity

The view of Being as an indivisible whole is supported by Einstein's theory of general relativity: for *everything that there is*, space, time, matter, and energy are no longer independent of each other (that is, they are not absolute), as was the case with Newtonian physics, but are intimately interwoven, affecting one another constantly. "Time and space and gravitation have no separate existence from matter."[4] Spacetime is a malleable continuum distorted by matter.

Yes, it is true that for the sake of practical calculations in physics we often isolate, in our mind, a phenomenon of interest by assuming that it is disconnected from the rest of nature (disconnected from the whole). For example, we study the gravitational interaction between the sun and the earth by neglecting the gravitational effects of the rest of the heavenly bodies. But as in the philosophy of Parmenides modern physics is about oneness, not isolation. And in reality all things in nature are part of the whole and are entangled far more intricately than the theory of relativity alone could discover.

Quantum Entanglement

One of the most fascinating consequences of quantum theory is the phenomenon of quantum entanglement. According to it, there are no perfectly isolated particles (or systems). The notion of an individual particle disconnected from the rest of the universe is inaccurate. Rather, all particles in the universe are part of a unified whole. They are in constant and *instant* interaction, affecting and determining the behavior of each other regardless of how far apart they are. Quantum theory suggests that everything that happens in the universe influences instantly everything else. In this sense the universe is indeed a Parmenidean indivisible whole. To explain this concept further we use the following thought experiment.

Suppose, for simplicity, that a mother particle could initially be at rest and with zero spin, and that later it decays into two daughter particles, A and B. To conserve momentum (linear and rotational) the daughter particles must take off away from each other as well as spin in opposite directions. In 1935, Einstein, with Boris Podolsky (1896–1966) and Nathan Rosen (1909–1995), argued through this thought experiment (which is known as the EPR, from the initials of their last names), that the daughter particles must have a fixed spin *since the moment of their creation*. To conserve rotational momentum one must spin clockwise, the other counterclockwise. Which particle spins in what direction is determined with a measurement. So if Alice measures that particle A spins clockwise, she is also certain that particle B must spin counterclockwise, as it is so confirmed when Bob measures it. Einstein's view is really the deterministic view of classical physics: that a particle has a fixed property *even before we measure it*.

But according to quantum theory, the spins of the particles A and B become fixed *only when an observation (a measurement) is made*. Until then, not only do we not know how the particles spin, but even worse and unlike Einstein's view, the particles *do not have a fixed spin*; each particle is assumed to spin simultaneously in both directions until a measurement is performed that will force them to take on a fixed spin—a peculiar concept, which is known as the Copenhagen interpretation of quantum theory.[5] It is this interpretation that Einstein found illogical and aimed to refute. And so did Erwin Schrödinger: to capture the peculiarity of the indeterminate spin state that particles A and B were assumed to be in before the act of measurement, he used a metaphor, the famous Schrödinger's cat. Briefly, he argued that according to the Copenhagen interpretation, until an actual observation is performed, a cat in a sealed opaque box, which also contains radioactive atoms with a chance to decay and spread poison, is both dead and alive at the same time. Namely, the state of existence of the cat before the observation is a mix of two possible outcomes because the status of the cat depends on the status of the radioactive atoms, which, per the Copenhagen interpretation, themselves are in a mixed state of two quantum probabilities, that of the decay outcome, which will kill the cat, plus that of the non-decay, which will preserve the cat. Only after opening the box and observing can the observer actually determine whether the cat is definitely either dead or alive. Before the observation, the cat is both dead and alive, in the Copenhagen interpretation. But according to classical physics even before opening the box, the cat is in a definite state of existence: it is either only dead or only alive.

So, according to the quantum view, the spin direction of each particle is fixed by the very act of measurement. For example, if Alice measures that particle A spins clockwise, *then and only then the spin of particle A becomes fixed* (contrary to Einstein's view, for which A would have been spinning clockwise since its creation); and, equally importantly, *then and only then the spin of particle B becomes fixed, too*, and it is counterclockwise (also contrary to Einstein's view for which B would have been spinning counterclockwise also since its creation). In general, measuring a property of particle A instantly forces a certain fixed property for particle B, even though particle B is not measured directly. This view, which is really the phenomenon of quantum entanglement, appeared absurd to Einstein because it meant, he argued, that the measurement of the spin of particle A affects and fixes *instantaneously* the spin of particle B, even when such measurement is performed while the particles are light-years apart and across the universe. This instantaneous "spooky action at a distance," as nicknamed by

Einstein, was not required by his analysis since according to it the particles had presumably fixed spin since their creation. How can such instantaneous influence exist, Einstein thought. How is it that measuring a property of one particle instantly affects and fixes an earlier indeterminate property of another particle? How is it that the very moment that the spin of particle A is measured, A communicates instantly how it spins to particle B so that particle B can spin opposite (to conserve momentum)? It is a strange type of communication that occurs faster than the speed of light, in fact instantaneously, and appears to violate one of the main principles of relativity, that information cannot be transferred with a speed faster than that of light because it would violate causality. This bizarreness caused Einstein to believe that quantum theory was not a complete theory of nature. Observing solely particle A would not in any way influence particle B, which is spatially separated from A, he thought.

But he was wrong. These opposite views appeared for a while to be part of the unverifiable realm of metaphysics. But in 1964 physicist John Bell (1928–1990) found a way to convert each point of view into an experimentally testable calculation, which is known as Bell's inequality.[6] The experimental verdict found Einstein's view false and favored the spooky action at a distance of quantum entanglement! Indeed, by measuring the properties of particle A we instantly affect the properties of particle B regardless of how far apart they are. And so, generally speaking, by measuring a property of one particle in a system, what we actually measure is a property of the whole system—which includes us, the observer, too—or, more precisely, of the entire universe. The universe is indeed an indivisible entangled whole. In the Copenhagen interpretation the observer is really part of what he observes—there is a mutual influence between observer and what is being observed. Whereas in classical physics the observer is thought of as an outsider separated from what he observes—there is no influence at all between observer and what is being observed. The classical-physics view of an observer is therefore like someone watching a movie—if the movie is nature, then an observer eating popcorn and drinking soda while watching has no influence on the movie plot (on nature). Whereas the quantum view of an observer is like someone being in the movie—his actions are *part* of the plot.

Of course this constant and instant interconnectivity between things in the universe, this quantum entanglement, exists not just when we curious observers of nature exercise our free will and decide to make a measurement but is an intrinsic property of nature. For just as in an act of measurement, for which we observers cause willingly particles to interact in order to satisfy our inquisitive mind—for example, photons are shined upon electrons to see where they are and how they spin—particles in nature are in constant interaction anyway (without us having to cause it at will), as if nature itself is constantly self-measuring (self-observing). Now, with self-measuring in mind, we have an additional reason to reinforce a previous conclusion, that, not only are the phenomena *observed* to occur discontinuously (as a result of the very act of observation, as argued in chapter 12), but the phenomena *themselves* might occur discontinuously even when *we* are not observing, for *nature* is—*self-observing*.

The whole universe experiences the phenomenon of quantum entanglement. If two particles have a chance to interact initially (that is, to become entangled like particles A and B that were created from the decay of the same mother particle), they continue to interact (they remain entangled) even when they are later separated. With this in mind, the entire universe may be considered an entangled whole (where everything in it is in constant and instant interaction with everything else, a perfect Parmenidean whole), for initially, according to the big bang, the entire universe was a mere microscopic size, possibly just a single point, where certainly everything was in close interaction and thus entangled with everything else, and so must then continue to be so today even with everything so far apart. The cosmic interconnectivity of mathematical nature anticipated by the Pythagoreans is now taking a concrete form through quantum entanglement.

In concluding this section, I would like to emphasize that information that travels faster than the speed of light is still impossible as stated in the theory of relativity. That is, while Alice's measurements of the properties of particle A influence instantaneously the properties of particle B, still information, that is, what Alice knows concerning the properties of either particle A or B, cannot be communicated to Bob faster than the speed of light; each person's knowledge can be communicated to one another at best at the speed of light, say, by a radio signal. Only then can Alice and Bob verify the remarkable correlation between the properties of particles A and B due to the phenomenon of quantum entanglement. Before such communication, the outcome of Bob's measurement concerning the spin of particle B would appear to him as random, as dictated by the laws of quantum probability, even when Bob does his measurements after Alice has done hers.

So as attested indirectly by the motion, change, and plurality of everyday experience—when these of course are investigated rationally by the human mind—the universe is indeed an indivisible whole. But there is a hypothesis that such universal oneness was once truly absolute.

THE ABSOLUTE ONENESS

The ultimate example of Parmenidean oneness, wholeness, and completeness, as properties of the universe, comes perhaps from the cosmological model of the big bang. It speculates that all matter and energy, all of space and time, the absolute wholeness of the universe of today, was once, about 13.8 billion years ago, contained in just a singular point. This primordial point, we must emphasize, was not within the universe; this one point *was* the universe, the *whole* universe; infinitely small, hot, and dense, containing a single type of particle and obeying one grand law—the absoluteness of oneness.

Our current big bang model begins its speculations on the universe's properties as early as the unimaginably small period of about 10^{-43} seconds after the initial bang—then the universe was extremely small but not of size zero. As the universe expanded and cooled down, its absolute oneness—the single primordial point that is hypothesized to have once been—and intrinsic simplicity manifested themselves as plurality and complexity, as particles of matter, the quarks and leptons, and as four groups of force-carrying particles, the photons, the gluons, the W's and Z's, and the gravitons. Quite possibly we will discover other types of exotic particles. The particles of force coalesced the particles of matter into nuclei, atoms, molecules, planets, stars, galaxies, books, and readers. But if the universe was once characterized by an absolute oneness, should it not continue to be characterized as such even today?

Unfortunately, the properties of the universe at this hypothetical primordial singular state cannot be described even in principle by our current physical theories. At this singularity—when the universe's size and age are both identically zero—all our physics equations break down; they are meaningless. Could this breakdown be an indication that such a singular state of the universe is really an impossibility, a Not-Being? If so, the universe might have been very small but not point-like. But was it born?

UNBORN AND IMPERISHABLE

Parmenides's philosophical worldview is, so he says, presented to him as a revelation by a goddess and is described in his poem *On Nature*. The main parts of the poem are the "Way of Truth"[7] (which discusses his philosophy) and the "Way of Opinion"[8] (which, among other things, discusses the philosophies of other philosophers). His primary goal was not so much to create a specific physical theory that would explain particular phenomena of nature but rather a theory attempting a logical explanation of existence itself: how can something be? It just is, he reasoned, for there is no such thing as nothing. Nature is unborn and imperishable. That which exists can neither be created from nonexistence nor obliterated into nonexistence. If the universe had a beginning, it would mean that it once did not exist—for if it existed it could not begin. But if the universe did not exist, it would have been Not-Being, and so again it could not begin (for Being cannot come from Not-Being). So the only way to explain why the universe is, is to assume that what is, has always been, unborn, without a beginning.

Now, on the one hand, his view of an unborn nature means that nature has not been caused; it does not have a primary cause. On the other hand, the opposite idea is that nature has been caused by a primary cause. This latter view is in a sense antiscientific since the premise of science is comprehensibility. But a primary cause cannot be understood—if it could, we would know what caused the "primary" cause, hence the "primary" cause would not have been really primary. Conversely, an unborn nature seems, at least at a first glance, to be more in accordance with the scientific premise, because something unborn/uncreated does not require a primary cause (an explanation) of why it exists— for it has always been.

That said, the notion of an unborn (uncreated, uncaused) nature (or, analogously, an imperishable nature having no ultimate purpose) must be examined with more caution. For it does not exclude the possibility of a god coexisting in the whole—as is in fact the case of the Parmenidean "Way of Truth," according to which the apocalyptic goddess Parmenides, and all the rest of nature, all just *are*. Moreover, an omnipotent and omniscient god could have made nature appear uncreated to us mere mortals. The point is that science cannot prove or disprove the existence of a god, and therefore such notion, as Parmenides might have put it in his "Way of Opinion," will always remain a matter of subjective belief. In science we must always begin with an assumed something (a Being)— and if we happen to finally explain such assumed something, we explain it with a new assumed something. Science cannot begin from Not-Being: *there is no scientific explanation of a universe coming to be from nothing!* Why there is something instead of nothing is scientifically unanswerable. Causality in our theories explains only later effects by earlier causes, but it cannot explain the primary cause (the beginning). And as a consequence, there is no way to ever know if there is an ultimate purpose. Even if the truth of the universe is revealed to us, the only way we can know that such truth is *absolute* is if we ourselves have absolute abilities—so that we can comprehend the absoluteness in the revealed truth. But we do not. And so again, the true nature of such hypothetical revelation is subjective.

Among our best cosmological models in science, the big bang does not and cannot answer why there is a universe; it only assumes that there is one (that might have begun or might have always been) and then continues to describe it. But it cannot answer why there is what there is. The prediction of the big bang model, that the age of the universe is 13.8 billion years, is only a *relative* age, namely, that our scientific theories can begin describing the properties of the universe roughly since 13.8 billion years ago. But we emphasize that with regard to what the universe might have been doing before that, we are clueless.

Interestingly, in an effort to avoid the breakdown of the equations at the hypothetical big bang singularity, some cosmological models attempt to model mathematically a self-reliant universe with neither space nor time boundaries—that is, an unborn nature having neither beginning nor end. A geometrical analogy of such type of universe is the surface of a sphere: no place on it can be considered more of a beginning than an end, more of a center than an edge, or more special in any way. Of course once more we emphasize that a mathematically/scientifically unborn universe does not say much about the true origin of the universe—that remains a subjective matter.

Lastly, the Parmenidean view is a hopeful philosophy because within its context consciousness is part of Being—I think, therefore I have consciousness. Hence, consciousness can never become Not-Being, even with the body's apparent death.

CONCLUSION

After Parmenides, any new natural philosophy would be considered incomplete unless it could address successfully his various conclusions, which, though unconventional, were logical. And as if that by itself was not a formidable task, Parmenides's best student, Zeno, assertively supports his teacher's views by adding to the complexity with his famous paradoxes that question the very nature of plurality, space, time, and the reality of apparent motion.

CHAPTER 14

ZENO AND MOTION

INTRODUCTION

Through a series of so-called paradoxes, Zeno of Elea (ca. 490–ca. 430 BCE) tried to argue for the astonishing conclusion that motion is impossible and plurality is an illusion. Could he be right? We present four of his most daring paradoxes: the dichotomy, the Achilles, the arrow, and the space, which challenge various views on space, time, and motion, and examine them within the context of modern physics. We also refer briefly to the conclusion of his paradoxes on plurality, which deal with whether there are many things or just one.

There is still no commonly accepted resolution for any of Zeno's paradoxes, a fact that preserves their legacy as the most difficult and long-standing puzzles. Part of the reason for this is the involvement of key notions such as space, time, and matter, of which their true nature is far from known even by the standards of modern physics. The real resolution of the paradoxes might require an even more radical understanding of these notions than the one presently provided by general relativity and quantum theory. Proposed solutions have often aimed to prove that motion is real. We will argue in favor of Zeno that at best, whether motion occurs or not is not experimentally provable.

THE DICHOTOMY PARADOX

According to Aristotle's account, Zeno said "Nothing moves because what is traveling must first reach the half-way point before it reaches the end."[1] In order to interpret this quote we must suppose that space is either (1) infinitely divisible (where space is imagined to be divided to ever smaller fractions) or (2) finitely divisible (where space cannot be divided beyond a fundamental length).

185

(1) Infinitely Divisible Space

The paradox can be interpreted two different ways, both of which are essentially the same. In the first interpretation, the question is this: can a traveler ever start a trip? To begin a trip of a certain distance a traveler must travel the first half of it, but before he does that he must travel half of the first half, and in fact half of that, ad infinitum. Since there will always exist a smaller first half to be traveled first, Zeno questions whether a traveler can ever even start a trip.

In mathematical language, the traveler will be able to start his trip only if he can first find the smallest fraction (the "last" term) from the following infinite sequence of fractions of the total distance: $1/2, 1/4, 1/8, 1/16, \ldots$. But such smallest fraction does not exist; it is indeterminable (in fact this is what is meant by calling such a sequence of fractions infinite). So the paradox is this: while on the one hand Zeno's argument, which questions the very ability to even start a trip, is logical; on the other hand, all around us we see things moving. Hence either Zeno's reasoning is wrong or what we see is false.

In the second interpretation, the paradox can be reformulated in a sort of reverse manner. In such case the question will be: assuming a traveler can somehow start a trip, can he ever finish it? To finish a trip of a certain distance a traveler must first travel half of it, then half of the remaining distance, then half of the new remaining distance, ad infinitum. Since there will always exist a smaller last half to be traveled last, Zeno questions whether a traveler can ever finish a trip.

"Answers"

First note that by getting up and walking, as Antisthenes the Cynic[2] did after listening to Zeno's presentation and thinking that a practical demonstration is stronger than any verbal argument, is not at all a refutation of Zeno's paradoxes of motion, because Zeno does not deny *apparent* motion; he questions its truth. The pre-Socratics were well aware of the deceptiveness in apparent reality; what we see happening is not necessarily happening the way we see it.

An "Answer" Based on Simple Mathematics

With the first interpretation in mind, to start the trip the traveler must first figure out the smallest fraction of the total distance, that is, the "last" term of the infinite sequence of numbers 1/2, 1/4, 1/8, 1/16, ... Only then he will know where to step first and begin the trip. But such term is indeterminable. After infinite subdivisions of the total distance, the "last" term of the sequence is indeed infinitesimally small and *approaches* zero, though is not *exactly* zero: there will always exist a smaller first half to be traveled first. Now since such term approaches zero we might want to approximate it to exactly zero. But with such approximation the traveler will step first where he is already at, the beginning. This might be interpreted to mean that the trip cannot start, thus motion is impossible. Nonetheless, this is not necessarily the best conclusion since it is reached only after our convenient approximation of the "last" term with the number zero. Since the actual value of the "last" term is indeterminable, a better conclusion would be that, indeterminable must also be the status of the trip (whether the trip can ever begin). Thus the notion of motion is, to say the least, ambiguous. The same result is obtained through similar arguments applied to the second interpretation of the paradox.

An "Answer" Based on Modern Mathematics

Often an answer of the paradox is sought through calculus. Suppose the trip distance is 1 meter. Then, as per interpretation two, a traveler first travels half of the trip distance, that is 1/2 of a meter, then half of the remaining distance, that is, an additional 1/4 of a meter, then half of the new remaining distance, that is, an additional 1/8 of a meter, ad infinitum. To find out if the traveler covers the trip distance of the one meter, we must add all the segments traveled by him, that is, 1/2 + 1/4 + 1/8 + 1/16 + ... Because the sum of this infinite geometric series converges on 1, some argue that the distance traveled by the traveler after infinite steps is 1 meter, thus he has moved and the paradox is resolved.

But this argument has a flaw hidden in the details of calculus. To be able to do calculus (i.e., calculate series sums like the one in hand) irrational numbers must be approximated with rational. Recall from chapter 11 that there exist infinitely many irrational numbers along any space distance. For example, between the point zero (the beginning of the trip distance) and the point of 1 meter (the end of the trip distance), there are infinitely many irrational numbers—such as $\sqrt{2} - 1 = 0.414213562\ldots$, or half of it, or one third of it, and so on—that must be approximated with rational numbers before any sum is calculated. For example, approximated to four decimal places, the rounded-off value of $\sqrt{2} - 1 = 0.4142$. Zeno, however, seems to tacitly question these very approximations that are required in mathematics to make the series convergent to a practical and calculable answer. Because nature, he would claim, does not have to behave according to the result of such convenient and ambitious human approximations.

Furthermore, some argue that the convergence method does not address the paradox because it does not explain how an infinite number of tasks (going from the first half of the distance to half of the remaining, etc., ad infinitum) can be carried out in finite time. But can a finitely divisible space solve the paradox?

(2) Finitely Divisible Space

In a finitely divisible space it is assumed that there exists a fundamental length that cannot be divided further. Therefore, there exists only a finite number of fractions of the total distance, and the paradox appears resolved, or, more precisely, the question of the paradox, as we will see, is revived in a new form.

For example, say space is finitely divisible and composed of fundamental lengths equal to 1/4 of a meter (which means that space cannot be divided in smaller lengths than 1/4 of a meter). How could a traveler complete a trip of distance 1 meter? Having in mind the first interpretation, the traveler steps first at the 1/4-meter location, then at the 1/2-meter, and finally at the 1-meter, the final destination. Analogously, by interpretation two, the traveler steps first at the 1/2-meter, then at the 3/4-meter, and finally at the 1-meter, the final destination. But a finitely divisible space generates a series of new unresolved questions. For example, how can the traveler move to the 1/4-meter location without passing first through all other locations (such as the 1/8-meter or the 1/16-meter, etc.)? Also, what determines the fundamental length of a finitely divisible space—whether 1/4 of a meter or some other number?

THE ACHILLES AND THE TORTOISE PARADOX

"In a race the faster runner can never overtake the slower. Since the faster runner must first reach the point from which the slower runner departed, the slower runner must always hold a lead" (Aristotle's account of Zeno).[3]

The paradox says that in a race between, say, fast Achilles and a slow tortoise, initially separated by some distance, Achilles can never overtake the tortoise because before he achieves that he must first reach the starting location of the tortoise. But by the time he arrives there the tortoise will have had the chance to move to a new location forward; and by the time he arrives at the tortoise's new location, the tortoise will move farther forward to another new location, ad infinitum. Therefore, despite that faster Achilles will be constantly approaching the slower tortoise, still there will always exist some small and ever-decreasing distance separating them (though not necessarily in fractions of half, as in the dichotomy paradox). This is a paradox because, despite Zeno's argument that a faster runner cannot overtake a slower one is logical, fast runners apparently do overtake slower ones. Is Zeno's reasoning flawed, or are our senses false?

This paradox is basically the same as that of dichotomy, so everything mentioned earlier applies here. One important difference is that the Achilles paradox is complicated further by contemplating the relative motion between two things (of Achilles and the tortoise), and between those things and their potential meeting point (destination); whereas the dichotomy paradox contemplates the relative motion of just one thing (of a traveler) with respect to a destination. Furthermore, the Achilles paradox contemplates the nature of time more directly (i.e., whether infinitely divisible, that is, continuous, or finitely divisible, that is, discrete), since it involves the time required by Achilles to reach the tortoise and how during such time the tortoise has the chance to move forward.

In the dichotomy paradox, the first interpretation (for which a traveler cannot start his trip) seems to deny motion more directly than the second interpretation (for which a traveler is assumed to move, although he cannot ever finish his trip). In the Achilles paradox, Achilles and the tortoise are assumed to be moving, but motion seems to not work in the conventional way, for the faster Achilles cannot overtake the slower tortoise. In the arrow paradox Zeno is even more audacious, for he directly denies motion by any interpretation. Reconstructing it reads as follows.[4]

THE ARROW PARADOX

An arrow is at rest whenever it is in a space equal to itself. A launched arrow goes through its flight one instant at a time. Since the arrow is in a space equal to itself each instant of the flight (just as it is when it is at rest), then the arrow must be at rest at each such instant as well. Since it is at rest at any one instant, it must be at rest for the entire duration of the flight. Hence the flight is apparent, not real; the arrow does not move.

This is a paradox since its conclusion is based on a logical argument that contradicts the apparent reality of sense perception according to which a flying arrow changes positions each moment of its flight and thus apparently moves. Again, is Zeno's logic flawed or are our senses?

I believe the formulation of the arrow paradox must have been triggered by a simple observation, that an object at rest occupies a space equal to its own size. A book, for example, resting on a desk occupies a space exactly equal to its own size. That said, I am not implying that such observation validates Zeno's conclusion that an arrow in apparent flight does not move. But could he be right? Could it be true that an apparently flying arrow is really motionless? Using quantum theory I will argue that at best, it is not provable whether the arrow is moving or not. Motion, in general, is an ambiguous concept.

MOTION IS AMBIGUOUS

While motion is part of apparent reality and is also the very premise of important theories of physics, on a fundamental level (i.e., concerning the motion of microscopic particles, to say the least) motion has not yet been experimentally proven, and in fact never can be! Therefore, motion is essentially a postulate inferred from sense-perceived experiences, but its truth is actually ambiguous. This is so because inherent in the Heisenberg uncertainty principle observations are disconnected, discrete events; consecutive observations have time and space gaps—we can observe only discontinuously (as seen in chapter 12). The concept of continuity in observation must be dismissed. It is a false habit of the mind created by the observations of daily phenomena—as of an arrow in flight (although, as explained in the section titled "Observations Are Disconnected Events" in chapter 12 and as will be reemphasized below, the arrow's apparent continuity of motion is an illusion due to its large mass that makes the time and space gaps between consecutive observations undetectably small). Now without the ability to observe continuously, motion not only *is observed* to be discontinuous but the very notion of motion *itself* becomes ambiguous. How so?

Motion occurs when during a time interval a particle (e.g., an electron) changes positions; a particle should be now here and later there in order to say that it moved. But since nature does not allow us to keep a particle under continuous observation and follow it in a path, and also since a particle is identical to all other particles of the same family (for example, all electrons are identical), it is impossible to determine whether, say, an electron observed in one position has moved there from another, or whether it is really one and the same electron with that observed in the previous position, regardless of their proximity. Since observations are disconnected and discrete events—with time (and space) gaps in between, during (and within) which we don't know what a particle might be doing—subsequent observations of identical particles might in fact be observations of two different particles belonging in the same family and not observations of one and the same particle that might have moved from one position (that of the first observation) to the next (that of a subsequent observation). Without the ability to determine experimentally whether a particle has changed position, its motion—and motion in general—is a questionable concept.

In summary, (1) without the ability to keep a particle under continuous observation (2) it is impossible to establish experimentally its identity, and therefore (3) it is also impossible to prove experimentally that it has moved.

Reinforcing this conclusion is the fact that, when we observe a microscopic particle all we see through a microscope is just a flash of light, and somewhere within it is the particle. But where exactly within it the particle is each moment of time, and whether it is at rest or in motion, are all indeterminable; while we do detect a particle, we detect it neither at rest nor in motion. Hence indeed, neither immobility nor motion are experimentally provable. Motion is ambiguous.

Trying to capture the peculiar consequences of the Heisenberg uncertainty principle concerning motion, physicist J. Robert Oppenheimer (1904–1967) wrote: "If we ask, for instance, whether the position of the electron remains the same, we must say 'no'; if we ask whether the electron's position changes with time, we must say 'no'; if we ask whether the electron is at rest, we must say 'no'; if we ask whether it is in motion, we must say 'no.'"[5]

Now, since motion is ambiguous for microscopic particles, then in a stringent sense it must be ambiguous for arrows, too, for arrows are made of microscopic particles. Observing an arrow in flight moving continuously does not prove (in the strictest sense of the word) that one and the same arrow has endured and moved, simply because there is no proof that any of its component microscopic particles have endured and moved. Besides, quantum theory (hence the uncertainty principle) is true for both the world of the large and the small. It is only for practical purposes that the world of the large is assumed to behave according to classical physics—for which objects appear to endure and move in definite traceable paths—because the consequences of the uncertainty principle for large objects are undetectably small, although not zero. Well, then, how do we explain everyday apparent motion and in fact the apparent continuity of apparent motion for any object, such as an arrow in flight, Achilles chasing a tortoise, or anyone taking a trip?

CINEMATOGRAPHY AND APPARENT MOTION

We can explain them with cinematography. In an analogy, consider a series of identical and disconnected red lightbulbs, closely spaced along the arched outline of George Washington Bridge. Now imagine it is nighttime and that the first lightbulb in the series is turned on briefly for a few seconds then off forever; after a brief time gap, lasting a minuscule fraction of a second, the second lightbulb is turned on and off the same way, then the third, and so on, until each lightbulb is turned on and off in this sequential manner. The events, the on-off turnings of each lightbulb, are (1) identical (in the sense that an observer sees the same red light) and (2) disconnected: the space gaps are the distances between the bulbs; the time gaps are the minuscule fractions of a second. Furthermore, (3) assuming that the space and time gaps of these events are small enough, to a distant observer, this phenomenon will appear as if *one red object* (the first lightbulb) has *moved* and has moved *continuously* along the outline of the bridge, when in fact no object has. Motion in this case is an illusion of the senses created by observing a series of identical and appropriately discon-nected events. In particular, the first two facts, (1) and (2), create the illusion of motion, and requirement (3) creates the illusion of continuity of motion.

The way the red light appears to be moving is similar to the way an arrow in flight appears to be moving. In each case apparent motion and apparent con-tinuity of motion are in reality the result of (1) a chain of identical observations (of an apparently identical red light or arrow), which (2) are also disconnected (for the arrow, this is due to the uncertainty principle) with (3) undetectably small space and time gaps in between (for the arrow, this is due to its large mass). Specifically, (1) and (2) create the illusion of motion, of one and the same object, the red light or the arrow, and (3) creates the illusion of *continuity* of motion.

But to refine the analogy, we must add this: unlike the case of the bridge for which several identical lightbulbs are assumed to preexist along its outline, for an arrow in flight we cannot assume several identical arrows to preexist along its apparent path; only that, each one of our *observations* is of an apparently identical arrow; though it is uncertain whether our observations are absolutely of one and the same arrow, for, as we learned in chapter 12, it is impossible to establish experimentally the identity of microscopic particles, and since arrows are made up of such particles, it is also impossible to prove unambiguously whether our observations are actually of one and the same arrow (that is, of an arrow composed of the *same* microscopic particles at each location of its apparent flight)—a fact that makes motion an ambiguous notion for any object, microscopic or macroscopic. The most we can say is that, at subsequent obser-vations the observed arrows have similar bulk properties, similar general *form* or *organization* (just as the Heraclitean river); it is this general organization that seems to endure, at least during some interval of time and within some region of space, creating the illusion of a permanent thing (e.g., an arrow) in motion.

While on the one hand all around us certain things appear to endure (appear as *permanent things* at least for some time and within some region of space), and whenever they appear to move there appears to be continuity in their motion, on the other hand, neither permanency in things nor motion are experimentally provable ideas. Thus motion is, to say the least, an ambiguous concept, a result to be expected because of the very definition of motion, which requires that permanent things exist so they can move: motion occurs when during a time interval a *particle* (a *thing* in general) changes positions; but to refer to a particle and define its motion, the particle must *remain the same for the duration of its motion*; motion cannot be defined if a particle does not remain the same for a period of time. Now for Heraclitus and modern physics there are no identifiable particles, no permanent things, only events. And without an enduring thing, without the ability to establish the sameness of a thing at two different moments, motion remains an ambiguous concept. While ambiguous, can it nonetheless be a practically useful way to explain the phenomena?

ADEQUACY VERSUS TRUTH

The answer to the previous questions is sometimes. Causality in classical physics is deterministic: a single cause produces a single effect, and both cause and effect are precisely determinable at least in principle. In quantum theory causality is probabilistic: causes and effects are expressed in terms of a probability; this is actually the reason it is impossible to determine whether the observations of two identical particles might be observations of one and the same particle, for we cannot causally connect these observations with deterministic (absolute) accuracy. Still, to make sense of the phenomena from the point of view of quantum theory, we often assume certain causal chains of events. For example, an electron here collides with a photon and recoils there (as if the electron endures). Thus while neither a cause nor an effect are certain, and motions are untraceable, still the assumption of a certain chain of events and of motion is often an adequate way to model a practical explanation. Supposing that particles endure, we have previously argued that they are constantly moving.

But while motion may be an *adequate* and useful concept in devising a certain practical explanation of nature—especially so for macroscopic objects such as arrows, cars, and planes—as a *true* property of nature it is, to say the least, an ambiguous concept, for it lacks the support of experimental confirmation from the microscopic constituents of matter that make up all macroscopic objects. Therefore, the merit of modeling an object (an electron or an arrow) as moving is a practical necessity of everyday life, not a confirmable truth. It should also be pointed out that even practicality cannot be applied consistently (especially so in the microscopic world).

Before quantum theory, and within the context of classical physics, concepts such as position, velocity, and motion (in general), were intuitive, self-evident, and could be used in a definitive way to characterize an object; an object moves with a specific velocity, it is now passing from here and will later go there. However, after quantum theory all these concepts became counterintuitive and could not be used in the same definitive way to describe the behavior of microscopic particles; such particles have neither definite position nor velocity nor a path of motion—as seen, this ambiguity of motion first came up in the natural philosophy of Zeno. A better way to describe the behavior of a particle (and in general the phenomena), then, is not through motion but through the probabilities of quantum theory. It is the concept of probability that is the fundamental (intrinsic) property of matter and not properties such as position, velocity, and motion. Within this context, as initially argued in chapter 11, a particle is truly a mathematical form. Could this mean something concerning change and motion in nature?

A COMPROMISE: YES TO CHANGE, NO TO MOTION

In the view of Zeno the arrow itself exists but does not move, and there is no change. On the other hand, in the view of Heraclitus, the arrow as a permanent thing does not exist (only events exist), but constant change and constant motion do; only the *organization* of the arrow exists and endures at least for some time. Could these antithetical views be reconciled by modern physics? Well, we can observe an electron (or an arrow) here now and an electron (or an arrow) there later. So obviously we can experimentally confirm that there is a certain change of events (at least in what we observe and where and when we observe it, that is, our observations are of different phenomena, electrons or arrows, at different places and times); but we cannot experimentally confirm that anything has moved. So the compromise might just be this: that constant changes do exist in nature (as Heraclitus posited), but motion does not (as Zeno theorized). But do these changes occur in a passive, playground-like space, or are space, time, and matter somehow related?

THE SPACE PARADOX

"If everything that exists is in some space, then that space, too, will exist in some other space, ad infinitum" (Aristotle's account of Zeno).[6] We may reconstruct this quote as follows.

(1) Things that exist do so in some space.
(2) Space exists (for if it didn't [1] wouldn't hold and things could not exist).
(3) Since space exists and everything that exists is in some space, then space, too, must exist in some other space, ad infinitum.

With this paradox Zeno seems to be arguing that requiring space, that is, void (as the atomists do, by treating it as a sort of a playground to put things in), is as problematic as denying space (as Parmenides does). And that, if we shouldn't completely deny space, we also shouldn't treat space as a playground—as if space is supposed to exist independently of the objects merely for the objects' sake, namely, for them to exist in it.

As discussed in chapter 11, the theory of special relativity replaced the playground-like space of Newtonian physics (for which space and time are absolute) with the spacetime continuum (for which space and time are relative). For relativity things do not just exist in a passive space with time flowing steadily in the background. Instead, space, time, and matter are complexly intertwined—with astonishing effects such as length contraction, time dilation, relativity of simultaneity, and space distortions (the latter being true only in the theory of general relativity).

So Zeno's space paradox is a paradox because while, on the one hand, his argument against a passive playground-like space is logical, on the other hand, it contradicts sense perception of exactly that kind of space. With the theory of relativity in mind, the space paradox may be considered resolved.

The peculiarity of a playground-like space implied by the space paradox is appreciated further when we try to construct a similar-type time paradox (though this is not one of Zeno's paradoxes). For example, if everything that exists does so for some time, then that time, too, will exist for some other time, ad infinitum. This time paradox argues against an absolute (Newtonian) time, flowing the same way for everyone while things happen.

Last, in his effort to show that a nature made up by many things is as problematic and contradictory as the Parmenidean oneness, Zeno devised several other paradoxes. Based on them he concluded that if in nature there are many things, they must simultaneously be (a) infinitely small and (b) infinitely huge and (c) finitely many and (d) infinitely many.[7] We will not cover these paradoxes here.

CONCLUSION

Zeno's paradoxes challenge our views on the very nature of space, time, and matter. Are these notions somehow connected? Is there just one primary substance of matter, or are there many? Is the nature of matter continuous—spread everywhere and also infinitely divisible for which matter can be cut to ever smaller pieces? Or is the nature of matter atomic—and so finitely divisible, for which matter cannot be cut beyond some fundamental pieces that are spread unconnectedly because they are surrounded by void? In 1916 Einstein addressed successfully the first question with his theory of general relativity, in which spacetime is a continuum in constant and intricate interaction with matter. Empedocles, Anaxagoras, and Democritus took up the other three. Matter is atomic for Democritus but continuous for Empedocles and Anaxagoras. And while plurality in the number of primary materials is possibly speculated first by Empedocles, all three philosophers had a unique take on it.

CHAPTER 15

EMPEDOCLES AND ELEMENTS

INTRODUCTION

Empedocles (ca. 495–ca. 435 BCE) managed to reconcile the antinomies between the Heraclitean becoming (the constant change) and the Parmenidean Being (the constancy) by introducing four unchangeable primary substances of matter: earth, water, air, and fire, later called elements, and two types of forces, love and strife. Change was produced when the opposite action of the forces mixed and separated the unchangeable elements in many different ways, an idea in basic agreement with modern chemistry or, more fundamentally, with the standard model of particle physics.

ELEMENTS AND FORCES

Unlike Thales, who taught that the primary material can transform and change its nature (for example, water can become ice), Empedocles held (as did Anaximenes) that the nature of a primary material must always remain the same, like the Parmenidean Being. But with a single primary material of unchangeable nature, he could not account for the observed material diversity of the world. Thus he postulated four such materials, the elements, which were uncreated and imperishable—neither born out of nothing nor perishable into nothing. His choice for these elements was wise because with them he could explain the three phases of matter: the element earth could account for the solid phase, water for the liquid, air for the gaseous. Furthermore, through fire he could account for light. Now, not only do the elements not change into one another; they do not change at all. But that did not matter. Because Empedocles explained nature's enormous diversity by imagining love (the force) mixing the elements with one another and strife (the other force) separating them from each other, in infinitely numerous proportions and combinations, forming composite objects or dismantling them. For example, love can mix earth and water to produce mud, but strife can separate the earth and water from mud. Hence love causes attraction of unlike elements (thus, in a sense, per Aristotle, indirectly it also causes repulsion of things that are alike). And strife causes repulsion of unlike elements (thus, in a sense, indirectly it also causes attraction of those that are alike).

Empedocles explained the unique properties of objects in terms of the proportions of the elements they contain. A hot object, for example, contained more fire than a cold object. And a wet object contained more water than a dry one. Thus, the quantitative difference of the various materials present in an object determines the qualitative difference between objects.

Birth and growth occur while the elements mix, as in a blooming flower, and decay and death occur while the elements separate, as in a shriveling flower. Coming to be (the birth, the generation of something) occurs simply from a mixture of things (the elements) that already exist, not from Not-Being (nothingness)—that is, there is no absolute birth. And perishing (the death of something) occurs simply from a separation into things that also already exist, not into Not-Being—that is, there is no absolute death. That there isn't absolute birth or absolute death is of course part of the Parmenidean thesis and is also accepted by Anaxagoras and Democritus.

Like the elements, the forces were corporeal, uncreated, unchangeable, and imperishable. But it was *their* motion through the elements that caused the elements to move, too—either pushing them together to mix or pulling them apart to separate. Hence, forces were the source of motion and consequently of change.

Force, in natural philosophy, appears for the first with Empedocles, who interprets nature in terms of matter and forces. Matter and force, however, became popular with Newton's work: first, with his three laws of motion, and second, with his law of the universal force of gravity. According to his second law of motion, for example, the cause of motion is a force: you pull an object and the object moves. Also matter can produce a force: the sun produces gravity, or an electron the electric force. Nonetheless, while the matter-force interpretation of nature is still immensely practical, it began fading away in twentieth-century physics: forces, in modern physics, gradually became no longer essential. This is a topic to be revisited in chapter 17. Force, in particular an action-at-a-distance type (as is Newton's force of gravity), we will see there, remarkably was never required in the atomic theory of Democritus.

EMPEDOCLES AND THE STANDARD MODEL

Empedocles's idea of forces mixing and separating a fixed number of primary materials is in fundamental agreement with the standard model of particle physics. Whereas Empedocles proposed two forces and four primary elements (renamed "particles" by physicist Leon Lederman),[1] the standard model considers three fundamental forces—the electromagnetic, the nuclear weak, and nuclear strong (recall gravity is not part of the standard model)—and twelve types of particles of matter—the six quarks and six leptons (even though various other considerations can increase the number to forty-eight: each quark comes in one of three "colors" [these are variations, not real color], and each matter particle has an antiparticle). Of course, unlike Empedocles's elements, in modern physics quarks and leptons are changeable—they transform to energy or from one material particle into another. Still quarks and leptons are brought together by the forces in a multitude of combinations and proportions to form atomic nuclei, atoms, molecules, flowers, and in general all the plethora of small and large objects, animate and inanimate, similar and dissimilar; but the forces can also break down larger objects into smaller ones.

In Empedocles's chemistry every object is made by a unique mixing proportion of the elements—for example, a bone, he says, is two parts earth, two parts water, and four parts fire (though the sources do not explain how he derived that). Analogously, in modern chemistry every chemical compound is made by a fixed proportion of the chemical atoms—for example, a water molecule, H_2O, is always made of two atoms of hydrogen and one of oxygen. Of course modern chemistry can be analyzed even more fundamentally within the context of the quarks and leptons of particle physics, and still preserve Empedocles's notion of fixed proportions. That is, H_2O, for example, is really a fixed mixture of two protons (one from each hydrogen nucleus) plus eight more protons as well as eight neutrons (from the oxygen nucleus) plus two electrons (one from each hydrogen) plus eight more electrons (from the oxygen). Now, electrons belong in the lepton family of particles, thus they are fundamental (they are not made of other types of particles), but protons and neutrons are not fundamental: a proton is made from three quarks, two up and one down ("up" and "down" are quark names); a neutron is made from one up and two down quarks. In addition, quarks and leptons are kept together or pushed apart via the continual exchange of force particles, the photons and gluons, in our example. Analogously, in Empedocles's theory the fixed proportions of the elements are achieved via the constant competition of love and strife.

Empedocles was interested not only in the composition and changes of individual objects but also of the world as a whole.

THE CYCLES OF THE WORLD

The structure of the cosmos is spherical for Empedocles, and the changes in it occur without an ultimate purpose or divine intervention (the latter is also the view of the atomists). Instead, nature is ruled by necessity and chance: namely, only some outcomes are possible (this is what is meant by necessity), but which ones actually occur is completely the result of chance. Interestingly, this is the meaning of probability in quantum theory.[2] Which outcomes (necessities) are possible and what the probability (chance) of their occurrence is are calculable by the mathematical laws of quantum theory. While these laws restrict the outcome of an experiment to any one of a group of possible ones (this is the element of necessity), the laws do not specify which one should occur (this is the element of chance). A hydrogen atom, for example, cannot really have any energy. When observed it has only one energy from a group of allowed ones (necessities). But while in one of the allowed energy states, a transition into another of the allowed energy states occurs by chance.

Nature in the cosmology of Empedocles goes through everlasting cycles of growth and decay, gradually and continuously, through four basic periods.[3] In the first period of the cycle love dominates totally but temporarily, mixing the elements completely. In the second period strife begins its influence, and so there is a gradual transition to partial mixing and separating. In the third period strife dominates totally but also temporarily, separating the elements from each other completely, so each, in its pure form, occupies a different region of space: one region of the universe is occupied only by earth, another only by water, another only by air, and the last one only by fire. In the fourth period love makes its gradual comeback, and so again there is a partial mixing and separating of the elements. Life (the evolution of plants and animals) and nature in general as we know it (with the sun, planets, stars) are happening during the second and fourth periods. The state of our cosmos is temporary for Empedocles, and it is gradually being succeeded by another. Interestingly, if we are not myopic in our comparisons, these four periods have several similarities with modern cosmological models of the universe.

CYCLES IN MODERN COSMOLOGY

According to the big bang model, initially everything was completely mixed together, space, time, matter, and energy (like Empedocles's first period). Life as we know it was then impossible because the universe was tiny and superhot, without stars or planets, just a super-dense mixture of tiny particles. The universe has since then been evolving, reaching its astronomical size and diverse state of today, with galaxies, stars, planets, and life (as in Empedocles's second period). Now, if, as speculated by various big bang models, the universe is "open," it will continue to expand forever, increasing its size so much that ultimately everything in it will be completely separated (as in Empedocles's third period). It will then be a cold, lifeless universe without planets or stars, only isolated tiny particles. But if, as also speculated by other models, the universe is "closed," then after it goes through a third period (a state of maximum, though not necessarily complete, separation of everything in it, during which stars might fade out and die), it will stop expanding and will begin contracting, resulting again in life-bearing partial mixing and separation (as in his fourth period). The fourth period is much like the second, for as matter is brought together in a shrinking universe, the particles coalesce again to form countless light-giving stars and life-sustaining planets to orbit them. But in a "closed" universe the contraction will continue until the crushing force of gravity ultimately collapses the universe in on itself, and brings once more everything completely together (the first period all over), causing a "big crunch" (the opposite of the big bang). If things are so, we live in an ever-changing universe going through endless cycles of big bangs, expansions, contractions, and big crunches. But we are not sure. Still, we could describe rather accurately the main events in the universe and when they occurred by starting from the first moment of the big bang until now.

COSMIC CALENDAR

In modern cosmology all events in the universe span 13.8 billion years in time. To gain a perspective of such time vastness we often employ a cosmic calendar. It is a metaphor by which 13.8 billion years, the estimated age of the universe since the big bang, are compressed into just one calendar year. The initial bang, the big bang, is supposed to have happened at precisely midnight, 00:00:00 (which, in the twenty-four-hour time notation, is the 0th hour, 0th minute, 0th second) on January 1, causing the expansion of the universe. What caused the bang is still unknown, although it is speculated to have been a kind of repulsive gravity that is predicted from the equations of general relativity. But what is known is that this expansion has been happening ever since and up until now, the last moment of December 31 at 24:00:00, increasing the size of the universe from an unimaginably small size initially, possibly point-like, to today's immensity. What banged (expanded, stretched)? Spacetime did and still does. Within a minuscule period of time after the initial bang, possibly by a mere 10^{-36} second, the universe underwent an immense faster-than-light expansion, a *big* bang, an idea known as cosmic inflation. In the blink of an eye it expanded by a factor of 10^{30}![4]

By about fourteen minutes (380,000 years) after the big bang, at 00:14:00 on January 1, the universe expanded, became less dense, and cooled significantly and as a result became transparent to light (as a clear-air day is to visible light), allowing for the first time the "afterglow" of the big bang, formally known as the cosmic microwave background, to travel freely through space and time, from there and then to here and now, and to be seen today (by radio telescopes) coming from every direction in the universe—a triumphal proof of the big bang theory. Earlier than the first fourteen minutes, the young universe was very dense and hot and thus opaque to light—as a foggy day is to visible light—thus light could not travel far. January 1, at 00:14:00, is also the instant that the simplest and lightest of the chemical atoms, hydrogen, first formed when a relatively cooler universe allowed electrons and protons to capture each other via the electric force.

Stars and galaxies began to form by around February 1 (about a billion years after the big bang) from matter pulled together by gravity. Stars shine because of nuclear fusion, the process via which light nuclei combine to form heavier ones, converting mass into energy and releasing light. Nuclei heavier than iron, including silver and gold, are synthesized via fusion when a supergiant star (more massive than the sun) becomes a supernova—dies violently in a cosmic explosion, producing as much light as a galaxy of stars!

Its death is also life's birth! For gradually after millions or billions of years, a supernova's scattered debris, an interstellar cloud of gas and dust, collapse again under the crushing force of gravity and grow into a new star with its orbiting planets that may also develop life. A perfect example is our own solar system. It was born much, much later, around September 3 (about 4.5 billion years ago) from the gravitational collapse of a massive interstellar cloud that was composed from the atoms that were synthesized earlier in the universe, including the heavy atoms made in the stars. Thus earth and everything on it, including us, are all made of these ancient atoms—if you are wearing a gold ring, you are in a sense actually wearing a portion of a star, for your jewelry's atoms were once manufactured in a supernova-destined star! Even more impressive, in the words of the great Carl Sagan, we are all "star stuff"! In other words, most of the atoms we are made of were once made inside stars that lived and died millions or billions of years before we or our own solar system were even born.

Primitive microscopic life-forms were thriving on earth by September 29 (3.5 billion years ago), so the first type of life must have evolved much earlier than that. On December 30 (sixty-five million years ago) an asteroid collided with the earth and caused the extinction of many species, including the dinosaurs. But that was a good day for primates because that's when they started to evolve. *Homo sapiens*, which are primates, evolved on the last hour of the last day of the cosmic calendar, December 31 at 23:52 (only eight minutes ago, two hundred thousand years ago). And at different moments during the last minute of the last day various other significant events occurred. Humans painted fine cave art one minute ago at 23:59 (thirty thousand years ago). They domesticated plants and other animals and gave birth to civilization twenty-three seconds ago at 23:59:37 (about ten thousand years ago).

Recorded history, which preceded the construction of the pyramids by a few centuries, began only eleven seconds ago at 23:59:49 (about five thousand years ago), and the birth of science occurred just six seconds ago, at 23:59:54 (2,600 years ago). Our innovative Internet was implemented about 0.08 seconds ago at 23:59:59:92 (in the 1980s), and a twenty-year-old reader of this book was born only 0.05 seconds ago at 23:59:59:95. If wisdom is, as the wise say, acquired with time, then human wisdom is only infinitesimal, not at all like that of the cosmos, infinitely universal.

What will a second such cosmic calendar be like for the universe? Will the universe continue to expand? Will it stop and begin to contract? We are not sure. While the cosmic microwave background together with Hubble's law constitute two of the most significant experimental proofs of the universe's expansion, an experimental proof concerning the universe's fate (if a particular one does exist) is yet to be found. Experiments are important because they verify or falsify a scientific hypothesis. Empedocles is known to have done an experiment, possibly the first in the history of science.

IT'S EXPERIMENT TIME

While air was the primary substance of matter in the philosophy of Anaximenes, still it was not accepted as a real corporeal substance for two reasons: (1) it is invisible and (2) because other objects appear that could be placed in air or move through it. So, within the context of these reasons, air was thought, at least by the Pythagoreans, to be really the void. But using a clepsydra (a device to lift and transfer liquids) Empedocles overturned such belief by experimentally proving that air is indeed a material substance.[5]

Submerge a straw (which is much like a clepsydra) in a glass of water. Water flows into the straw through its bottom opening and fills it as high as is the water level in the glass. But if before you submerge the straw you first cover its top opening with your finger, no water (or, actually, very little) will flow into the straw. This happens, Empedocles argued, because some invisible material, which is already trapped in the straw, presses on the water (through the bottom opening of the straw) and keeps it out; water, in this case, cannot move through this material. This material is air. Only when you uncover the top opening can water once again flow in the straw. For in this case the air in the straw escapes through the top opening, and so an equal volume of water flows in to take its place. (Incidentally, why water or air or any object can move will occupy the mind of the atomists, as will be seen in chapter 17). In conclusion, since it is not always true that an object can move through air or be placed in it, air must be a material substance, regardless of its invisibility. Empedocles's reasoning is correct.

THE ORIGIN AND EVOLUTION OF THE SPECIES

In his effort to understand the origin of the species and their adaptation to their environment Empedocles, like Anaximander, conceived of an evolutionary theory by natural selection. In the beginning a chancy mix of the "immortal"[6] (permanent, unchangeable) elements created all imaginable "mortal"[7] (temporary) organic "forms, a wonder to behold."[8] These, though, were just parts, from humans, animals, and plants. And so "many heads sprouted without necks, and arms wandered bare and bereft of shoulders, and eyes strayed up and down in need of foreheads."[9] That is, until love mixed them in countless ways more so that the species of plants and animals formed. But only the fit survived; the unfit died. When a human head, for example, combined with a human body, the creature acquired a fitting form and survived, Empedocles thought; but when a human head combined with an ox body, he continued, the creature was unfit and died. Chancy material combinations and natural selection (that is, survival of the fittest and adaptation) are important aspects in both Empedocles's and modern theories of biological evolution.

CONCLUSION

Empedocles's pluralistic philosophy was a crucial turn away from the monistic philosophies we have discussed so far (i.e., those that considered water, the infinite, or air as the only primary substance of matter, or the philosophy of Parmenides about oneness), for it paved the way for the most successful ancient pluralistic philosophy, the atomic theory of Leucippus and Democritus. Their theory required myriad particles: the atoms. But before the theory of atoms, pre-Socratic philosophy had to go through yet another theory of remarkable originality; four primary substances of matter for Empedocles, but infinitely many for the nous of Anaxagoras.

CHAPTER 16

ANAXAGORAS AND NOUS

INTRODUCTION

"Nous [the mind] set everything in order,"[1] thus it has the ability to understand nature rationally, Anaxagoras (ca. 500–ca. 428 BCE) proposed. Order though, according to him, is not achieved through the consideration of just one primary substance or even four but through a countless number of them, including things such as gold, copper, water, air, fire, wheat, hair, blood, bones, and in general all other existing substances. However, unlike Empedocles's four elements, which are pure, Anaxagoras's substances are not; "in everything there is a portion of everything,"[2] a notion as bizarre as two of the most popular interpretations of quantum theory, the Copenhagen and the many-worlds.

IN EVERYTHING THERE IS A PORTION OF EVERYTHING

All Materials Simultaneously

Every piece of substance, however large or small, contains some portion of everything—portions can be large but infinitesimally small, too, because for Anaxagoras matter is infinitely cuttable. Hence, no one substance is more fundamental (that is, smaller, simpler, purer) than any other. But "each thing is most manifestly those things of which it has the most."[3] A piece of gold, for example, contains gold as well as everything else—copper, wheat, hair—but appears as a distinct golden object because its gold content is the greatest. This does not mean, however, that this golden object contains the substances in pure form, side by side, separated, and identifiable, and the amount of pure gold in it happens to be more. No! To the contrary, no matter how small a bit we may cut from such golden object, it will still contain a portion of everything—it will never be pure gold. Therefore, despite that this is a golden object, *every part* of the object is also *simultaneously* watery, woody, milky, bloody, bony, hairy, and every other material, but not just that—it gets stranger.

All Qualities Simultaneously

A strict interpretation of "in everything there is a portion of everything" means that an object has not only a portion of each type of substance but also a portion of all opposite qualities. In fact, according to some scholars, it is not really necessary to speak of the substances separately from the qualities because the qualities determine the type of a substance anyway.[4] Now, as with the substances, these qualities are not to be assumed to be side by side in an object or separated somehow, as if, say, an object has its right side wet and its left dry. Rather, "Things in this one cosmos are not separated from one another, nor are they split apart with an axe, neither the hot from the cold nor the cold from the hot."[5] So every part of an object is all the qualities simultaneously. For example, something hot is to some degree also cold. Or white snow, Anaxagoras argued, is to some degree simultaneously black, too—a statement of the same unusual meaning as Schrödinger's cat being simultaneously both dead and alive.

ANAXAGORAS AND THE COPENHAGEN INTERPRETATION

So for Anaxagoras an object is simultaneously hot, cold, wet, dry, hard, soft, sweet, sour, black, white, bright, dark, dense, rare, dead, alive, spinning clockwise, spinning counterclockwise, and all other opposite qualities. This is a peculiar interpretation of nature, for before we observe an object, the most we can say about the state of its existence is that it is a mix of all possible outcomes—of all opposite qualities though each with a different degree (portion) of contribution. Only after we observe the object can we describe it in a specific way, in terms of "those things of which it has the most," say, as golden, yellow, dry, cold, and heavy.

Remarkably, such interpretation is similar to the most popular interpretation of quantum theory, the Copenhagen view. According to it, before we observe something, the state of its existence is a mix of all possible outcomes (qualities), each of which has its own quantum probability to actually occur. If the idea of portion in Anaxagoras's theory is roughly associated with the idea of probability in quantum theory, then indeed, "in everything [a system of interest] there is a portion [is described by the quantum probabilities] of everything [of every possible outcome]." Recall how before an observation Schrödinger's cat is simultaneously both dead and alive (or how an electron spins simultaneously both clockwise and counterclockwise). And each of these potential outcomes has its own probability to actually happen. Only after we observe, the Copenhagen interpretation states, can we determine whether the cat is definitely either dead or alive (or whether the electron spins definitely in the one or the other direction), and in general, whether an object is, as Anaxagoras states, definitely golden, yellow, dry, cold, and heavy.

Now the reason Anaxagoras required that various portions of all qualities had to coexist simultaneously everywhere within an object and at all times is that he wanted to remain in accordance with the Parmenidean thesis, that Not-Being does not generate Being, and that Being does not become Not-Being. Something must always exist if it is to be observed, the thesis says. That is, if a quality were not already present everywhere in an object always, it could not have come to be later; because if it did come to be later, it would mean that Being could be generated from Not-Being, but that's impossible. Hence a hot object, for example, has to contain simultaneously both hotness *and* coldness everywhere within it and always, though in different portions. For if a hot object did not contain coldness, too, coldness would have been Not-Being (at least for that object), and therefore it could have never come to be (coldness could have never become a reality, a part of Being)—it would then be impossible for the hot object to be cooled down.

This concept has a certain similarity with the Copenhagen interpretation but also a certain difference. Concerning the similarity, the reason we may observe the cat to be alive (or the electron to spin clockwise) is that the cat's (or the electron's) state of existence before the observation is a mix of all possible outcomes, that is, a mix that includes a portion (the quantum probability of occurrence) of the alive quality (or the clockwise spin) *together* with a portion of the dead quality (or the counterclockwise spin). In quantum theory this mix state is expressed mathematically. And the outcome with the highest probability (portion, in the language of Anaxagoras) is the one most likely to be observed.

Analogously, the reason we may observe, say, hotness, in Anaxagoras's view of our previous example, is that the object's state of existence before the observation is a mix that includes a portion of hotness and a portion of coldness, but with the hotness portion being the highest.

But Anaxagoras is even bolder than the Copenhagen interpretation, a fact that brings me to their difference. He insists that the notion of the simultaneous existence of all qualities is true all the time, even after an observation. Hence in Anaxagoras's view, the cat is still both dead and alive (or the electron still spins in both directions, or the object is both hot and cold) even after we observe the cat to be only alive (or the electron to spin only clockwise, or the object to be only hot). But in the Copenhagen view, after an observation the cat is only alive (or the electron spins only clockwise, or the object is only hot). Thus, although we observe only one quality, and so the cat appears only alive (or the electron is detected to spin only clockwise, or the object to be only hot), for Anaxagoras the other qualities never cease to exist; he insists on this because he does not want to violate the Parmenidean thesis that if a quality ceased to exist, it would mean that that part of Being became Not-Being. Now, can Anaxagoras be somehow right on this, too? Can the cat somehow be both dead and alive even after we observe it to be only alive?

ANAXAGORAS AND THE MANY-WORLDS INTERPRETATION

Fascinatingly yes! According to the second-most popular interpretation of quantum theory, the many-worlds view, even though *we* observe the cat to be alive (or the electron to spin clockwise), in *another universe* (world, reality) the cat is dead (or the electron spins counterclockwise)! That is, an outcome that is possible but does not occur in our universe still occurs in another universe. In general, every outcome that could have occurred in our present reality (universe) but did not, branches off as an alternative reality (it gets realized) in a parallel (i.e., separate) universe; each parallel universe thus has its own unique reality that consists of events that could have happened in our universe but did not.

Hence, the many-worlds view is in closer agreement than the Copenhagen view with both Anaxagoras's theory as well as the Parmenidean thesis. For, Parmenidean Being (being everything that there is) can easily be interpreted to include every possible outcome of an observation. Then, in the view of many-worlds, it is not only before an observation that all possible outcomes coexist in a mix and are part of Being (as is also required by the view of Anaxagoras), but all such outcomes, in a sense, continue to coexist and are thus still part of Being even after an observation (also as required by the view of Anaxagoras); for each possible outcome occurs in its own parallel universe even when such outcome is not observed to occur in our own universe. Whereas on the other hand, in the view of the Copenhagen approach, though before an observation all possible outcomes coexist in a mix and are part of Being (as also required by the view of Anaxagoras), after an observation only what is observed to occur continues to exist (to be part of Being), and what is not observed no longer exists, as if part of what once existed, part of Being, became Not-Being (a situation in clear violation of both the view of Anaxagoras and the thesis of Parmenides). With the Parmenidean thesis in mind one might then say that the many-worlds interpretation of quantum theory is more accurate than the Copenhagen.

THE PERFECT "IN EVERYTHING . . . IS . . . EVERYTHING"

The ultimate example of "in everything there is a portion of everything" is the big bang singularity, the hypothesis that the primordial state of the universe was once, 13.8 billion years ago, a mere point. "In everything"—in the singularity, which *itself* was everything that existed, the whole universe—"there" was "a portion of everything"—matter, energy, space, time, *and* the laws (or ultimate law) they obey. And the reason the universe is diverse, with planets, stars, people, plants, is that, as Anaxagoras might have explained, there is only a *portion* of everything in everything and "each thing is most manifestly those things of which it has the most."

Now, what remains enigmatic is this: (1) if indeed the notion of "in everything . . . is . . . everything" was true at the singularity, why would it not be true always? And (2) can plurality, all the beautiful and diverse nature of today, unfold from an absolute oneness, the singularity? Both are open questions.

On the first question, Anaxagoras would have answered "in everything . . . is . . . everything" always and everywhere. On the second question, all three pluralists, Anaxagoras, Empedocles, and Democritus, believed that plurality must be absolute; that is, neither could plurality (the many) have come to be from an initial singularity (from what is initially one), nor a singularity (the one) from an initial plurality (from what is initially many). Nevertheless, whether nature is truly monistic or pluralistic is a question that has yet to be answered. The singularity hypothesis is problematic (the physics equations are meaningless at such state of existence), and, as we saw in the section titled "Unborn and Imperishable" (in chapter 13), some cosmological models try to avoid it. Will our nous ever know the nature of nature?

NOUS

Nous for Anaxagoras is the only thing that is truly pure, containing nothing else except itself. Only living things have nous. Nous is infinite, timeless, has all knowledge about everything, and is the primary cause of motion and thus of all unfolding physical phenomena.

I believe that Anaxagoras did not mean literally and in a mechanical way that nous can cause motion, change, and "set all things in order." Rather, he meant that our nous is capable of a scientifically logical explanation of nature, such as one that assumes motion. And in general that any model of nature our nous conceives is adequate, provided that with it the phenomena are set in order and explained *scientifically* (rationally). This might be what matters after all, especially if an absolute knowledge about nature, a much-pondered topic since antiquity, is an impossibility, for example, because the truly first and last causes of the universe are indeterminable. Of course if reality is one (that is, objective), so ultimately should the scientific model be that our nous will conceive to explain it. And since for Anaxagoras "the phenomena are a sight of the unseen,"[6] that is, what we see contains subtle information about things we cannot directly see, along with reason (nous), Anaxagoras says the senses also have a catalytic role in our attempts to set things in order. However, what is the origin of our ability to reason? Why do humans have an intelligent nous (mind), the most intelligent, in fact, of all the species that we know?

FROM WALKING TO THINKING

Anaxagoras believed that the cause of human superiority over animals is the hand. Although other primates walk upright occasionally, only humans walk upright habitually, and as a result only humans have freed their hands permanently and have been using them consistently. This unique trait of ours has been significant for both our biological and intellectual evolution because when Lucy, a species of the genus *Australopithecus*, about 3.2 million years ago "decided" to walk upright more habitually, it meant that two of her four legs were starting to evolve into hands, increasing her potential to use and make tools. Toolmaking stimulates thinking (silent and out loud, thus speech, too), which in turn refines toolmaking, which stimulates further thinking, in a continuous cycle, ultimately advancing both technology and the intellect, and making Lucy's distant relative, the *Homo sapiens* (us), indeed *sapiens* intellectually superior to any other animal (at least on earth). And thus the origin of such superiority might truly be the hands. What actually ignited this development was a purely chance mutation in the spine that allowed our hominid ancestors to stand upright and evolve their forelegs into hands, become environmentally fitter, and get naturally selected further.

Upright posture (and consequently free hands), speech, and a complex brain (nous) are among the most unique traits of the human species and the source of our resourcefulness. The brain is the center of operations. Speech is controlled by the left side of the brain, but the coordination of the movements of our hands comes from the back part of the organ. Such coordination took literally millions of years of evolution to be mastered, during which the brain was driving the hands, which in turn were driving the brain, causing the enhancement of both organs, developing each to its present advanced stage of evolution compared to the hands and brains of other species. Without this type of evolution we would not have been able to make our first tools; stack up stones; build homes, the pyramids, the Parthenon, the Empire State Building; or create cell phones, computers, spaceships, MRIs, or the LASER, but also we would not have been able to pursue other more abstract endeavors, such as religion, philosophy, science, and the arts.

At the same time, however, I wonder if there is a limit to such brain-hands interactive enhancement. Even worse, I wonder, what the risks today are from the plentiful technology made by our hands as a result of the ingenious science conceived by our brains (nous). Do we get to depend more and more on machines to think for us and on pills to save us, risking the weakening of both the mind and the body? If yes, our natural abilities might atrophy and our evolution might be stalled. We might even devolve, for habits have, as seen, a say on whether we get to evolve.

The name of Galileo is often associated with the first major conflict between religion and science. He was tried by the Inquisition of the Roman Catholic Church for his support of the heliocentric model, which the Church considered in contradiction of biblical accounts of an immobile earth in the center of the universe (i.e., the geocentric model). But it was really the science of Anaxagoras that caused such conflict first. He was charged with blasphemy by the Athenian democracy for thinking that "the sun is a fiery stone"[7] and not a god. He was tried and found guilty. Although he was defended by his student Pericles, the famous statesman, still by one account he was exiled, by another he was sentenced to death. Anaxagoras was an original thinker. He is credited with explaining eclipses correctly and for introducing philosophy to the Athenians. When asked why he was born he replied, "To theorize about the sun, moon, and heaven."[8] When told "You are deprived of the Athenians," he replied, "No, they are deprived of me."[9]

CONCLUSION

Whether the nature of matter is infinitely cuttable (without smallest pieces) or finitely cuttable (with ultimate smallest pieces that make up everything) is still an open question. Anaxagoras held the former, but Leucippus and Democritus held the latter: namely, matter is atomic and thus made up of disconnected, indivisible pieces known as the atoms and surrounded by empty space. What revolutionized science was the atomic theory of matter, an idea that is two and a half millennia old.

CHAPTER 17

DEMOCRITUS AND ATOMS

INTRODUCTION

Perhaps the greatest scientific achievement of antiquity, possibly of all time, was the realization of the atomic nature of matter. "There are but atoms and the void" Democritus (ca. 460–ca. 370 BCE) proposed.[1] And he understood the great diversity of material objects as complex aggregations of uncuttable atoms, the building blocks of matter, moving in the void, the empty space between them. Leucippus, who flourished between 440 and 430 BCE, invented the atomic theory, and Democritus, a true polymath and a prolific philosopher, developed it extensively. Uncuttable (the actual meaning of *atom* in Greek) are also the modern elementary particles of matter, the quarks and leptons, and although void is a controversial concept still, a kind of void is required to explain nature.

ATOMS

Ancient Atoms

Atoms, in the ancient atomic theory, are the uncuttable smallest pieces of matter, disconnected from each other because they are surrounded by void, space devoid of matter that was required to enable the atoms to move. Atoms are invisible, impenetrable, solid (absolutely rigid), indestructible, eternal, unchangeable, unborn (not generated by something else more fundamental), and imperishable (they do not transform into something else more fundamental). Atoms are therefore like many Parmenidean Beings. Unlike the elements of Empedocles, which represented four different types of known materials, or those of Anaxagoras, which represented infinitely many types, all atoms are made from one and the same type of material (although not from any particular one of the everyday, such as water or air). Atoms have no internal structure (they are homogeneous) but differ from each other only as regards their size and shape. Their only behavior is motion. Roaming around in the void of an infinite space there exist infinitely many atoms of various shapes: angular, concave, convex, smooth, rough, round, sharp, and so on.

Their motion is perpetual (so, then, is change in accordance to the Heraclitean worldview). Motion continues by itself without requiring a force (or a causal agent in general). The fact that constant motion continues by itself was discovered via experimentation first by Galileo and was later restated by Newton through his first law of motion, also known as the law of inertia (a body will remain at rest or in uniform motion until affected by a force). Atomic motion is also thought random because space was correctly assumed to be isotropic (having no special location or direction, i.e., space has no absolute up, down, right, left, in, out, center, or edge). Hence, when left to themselves (undisturbed), atoms had no reason to move more one way than another—all directions of motion were equally probable. Motion was not explained by Leucippus and Democritus; it was simply postulated to have always been, without a beginning. In fact atoms and the void were also postulated. Not accounting for the cause of motion received Aristotle's intense criticism, even though it is actually a normal scientific procedure. For we need to remember that postulating the truth of a certain beginning and proceeding from there to understand nature in a causal and rational way is the only way to do science. Science must begin from something (a postulate, an axiom, a primary cause); it cannot begin from nothing (Not-Being). For Democritus the atoms, the void, and motion were part of a primary cause, which by definition requires no cause of its own.

As they move, atoms collide with each other, bounce, rotate, some hook together (whenever their shapes are complementary) and assemble in a multitude of arrangements, forming all kinds of macroscopic (compound) objects that appear "as water or fire, plant or man,"[2] or unhook and disassemble, deforming (destroying) the objects. As atoms aggregate objects form and increase in size, and as atoms segregate objects change form and decrease in size (i.e., they perish). Just as the words "tragedy" and "comedy" are formed when letters (which can be thought of as atoms) from the same alphabet are combined differently, Aristotle had explained, the immense plethora of diverse objects is formed when atoms are arranged in space differently through their motion.[3] In fact, atoms were required by Leucippus and Democritus to explain exactly this diversity in nature: "For from what is truly one a plurality could not come to be, nor from what is truly many a unity, but this is impossible."[4]

Atoms have none of the conventional properties of matter such as color, taste, smell, sound, temperature, or even weight. The proof of this, Democritus thought, is in the fact that various objects are perceived differently by different people. Something sweet to me might be spicy to you. "To some honey tastes sweet to others bitter but it is neither," said Democritus.[5] So he explained the conventional properties in terms of the shape of the atoms and their motion in the void; shape and motion cause unique macroscopic arrangements (in objects, in people) and thus unique conventional properties. For example, the atoms of a hard object are more closely packed with less empty space (void) between them than the atoms of a soft object. Now, since the atoms of a soft object have more void to roam around they can be pushed there more easily. Hence, such objects feel squeezable and soft. Metals are made of atoms with hooks that hold them firmly interlocked, but liquids are made of round atoms so they can flow by each other easily. Sweet objects are composed of round good-sized atoms; bitter of round, smooth, crooked, and small; acid of sharp (so they can sting the tongue) and angular in body, bent, fine; oily of fine, round, and small. Even light was made of atoms (particles)—incidentally, Einstein won the Nobel Prize in Physics by interpreting light in terms of discrete particles: the photons. Black, white, red, and yellow were considered primary colors and associated with different shapes and arrangements of atoms.[6] Combinations of these four colors were in turn used to account for all color variations. Democritus worked out a detail theory on sensation. In general he argued that the constant motion of the atoms, which persists even when a composite object is seemingly at rest, causes some of them to be emitted by the object. Flying through the void, these atoms in turn ultimately collide with the atoms of a sense organ and create a unique sensation (a flavor, smell, color). By the way, a collision (recall section "The Uncertainty Principle" in chapter 12) is the first out of two steps/events that generate the famous Heisenberg uncertainty principle. Because Democritus realized the uncertain nature of atomic collisions (due to the random atomic motion), he argued that the knowledge we acquire by sense perception (e.g., observation) is "bastard,"[7] uncertain—recall that the Heisenberg uncertainty principle also states that the act of observation causes uncertainty.

All changes of the apparent world of sense perception, animate and inanimate, were reduced to the irreducible atoms and their motion in the void. This scientific reductionism is ambitious. In principle, it is also the goal of the modern theory of the elementary particles of matter, the quarks and leptons. But while there are striking similarities between these modern particles of matter and the ancient atoms, there are also serious differences. Only for purposes of comparison let us call the ancient atoms Democritean or D-atoms, and today's building blocks of matter, the quarks and leptons, QL-atoms.

D-atoms and QL-atoms: Similarities and Differences

Before we proceed with this comparison let us first briefly summarize the historic developments in search of the D-atom, the ultimate uncuttable piece of matter. Until about the end of the nineteenth century chemical atoms, such as hydrogen, carbon, oxygen, and so on, of the periodic table of chemistry, were thought to be the fundamental particles of matter, the D-atoms. But this idea turned out to be incorrect when the structure of the chemical atom was probed further and it was found that it was made up of electrons and a nucleus. In 1897 physicist J. J. Thomson (1856–1940) discovered the electron, and in 1908 his student physicist Ernest Rutherford (1871–1937) discovered the nucleus. Electrons (one of the six types of leptons) are still thought indivisible, but atomic nuclei not so; the latter are made from divisible protons and neutrons, though they are themselves composed of indivisible quarks. Six types of quarks were postulated to exist as elementary particles in the 1960s, and all six types had been discovered by the end of the twentieth century. Hence, chemical atoms are not fundamental; they have substructure and in fact are made of the QL-atoms, which are among the particles of the standard model of physics. Like the shadows, which *were* real but were not the real *objects* in Plato's parable of the cave, chemical atoms *are* real but are not the real *fundamental particles* (the smallest cuts of matter as envisioned by Leucippus and Democritus)—although the name "atom" has been undeservingly stuck on them.

On the other hand, both D- and QL-atoms are fundamental because they are not made from other particles; they are disconnected pieces of matter, indivisible (uncuttable), invisible, and the smallest, and their various combinations make up all material things in the universe. Neither the D-atoms nor the QL-atoms have any of the conventional properties of composite objects. These properties are really a consequence of the collective behavior of the D- and QL-atoms that make up these objects.

D-atoms are unchangeable, they do not transform, but QL-atoms do; they transform from one type of material particle to another and also into and from energy (although they do not transform into something more fundamental). But like matter, energy also comes as discrete bundles, as particles (e.g., photons), and so Leucippus's and Democritus's notion of discreteness as a property of nature is preserved. Furthermore, like D-atoms, which are made of the same substance, QL-atoms are made of the same substance, too, mass and energy (which are equivalent as per special relativity). And since D-atoms are indestructible, so is their substance, but so is the substance of QL-atoms, for the total amount of mass-energy in the universe is constant (as per the law of conservation of mass-energy). So the substance of both, the ancient and modern atoms, endures, while nature is constantly changing.

D-atoms have shapes and thus have nonzero size, QL-atoms are considered point-like, thus shapeless and size-less. There is both a challenge and a simplicity associated with each view. In the D-atoms we need to imagine all sorts of complex atomic shapes, but we need not worry about forces. Democritus did not introduce any (a topic to be revisited below). D-atoms, he explained, coalesce into composite objects as a result of their perpetual motion and complementary shapes. Also, composite objects have size since they are made of D-atoms, which themselves have size. On the other hand, QL-atoms lack shape and size, but they still combine. They do so via the exchange of the particles of force (the photons, the W's, the Z's, the gluons, also the gravitons if we find them). Being point-like thus shapeless is in a sense a simplicity, for it means that QL-atoms are internally structureless like the D-atoms. But it is also a complexity, for how can something of zero size, of zero extension in space, have properties such as mass, electric charge, energy, spin, and so on, and, even worse, how can composite objects have size and extension in space when their constituents do not? Within the context of the Parmenidean theory the "size" question may be restated as follows: how can size come from not-size? That is, how can Being (the nonzero size of macroscopic objects) come to be from Not-Being (from zero-size constituents)? This was impossible for Parmenides and Anaxagoras—that's why the latter, recall, posited "in everything . . . is . . . everything." Democritus solved the size challenge by postulating that matter cannot be divisible (cuttable) at infinitum; it must be finitely divisible with the smallest cuts to be the indivisible, the uncuttable D-atoms of nonzero size. For only then, he thought, could composite objects have size, if they are composed of things that themselves have size. Interestingly, the latest hypotheses for fundamental particles of nature include particles that *do* have size: as seen in chapter 8, these are one-dimensional strings or two-dimensional membranes. If these turn out to exist, we may not need to worry about size (Being) coming to be from not-size (Not-Being), an idea that would be pleasing to all three philosophers, Parmenides, Anaxagoras, and Democritus. The size challenge is revisited in the section on "Void or Not?"

D-atoms are postulated to move in order to comply with the apparent world and explain change. On the other hand, since motion is an ambiguous concept from the modern point of view (as seen in chapter 14), QL-atoms are postulated to move only as an *adequate* way of understanding the phenomena. Furthermore, QL-atoms are best regarded as events (as seen in chapter 12) and not as permanent Parmenidean Being-like entities (as are the D-atoms). QL-atoms' existence (their properties and behavior) is best described in terms of the quantum probability, a number that expresses only a potential event: for example, which type of QL-atom might be observed, where, and with which properties and behavior. Within the context of quantum theory, material particles (the QL-atoms) are less materialistic in the sense that they no longer have key properties that material particles were once thought to have in order to be called material: they are neither permanent, nor indestructible, nor unchangeable, nor deterministic, nor have they well-defined shapes or trajectories through space and time, and as a consequence nor have they identity and individuality. They are represented by quantum probabilities, thus, as has also been argued in chapter 11, they are mathematical forms. These mathematical forms have often been correlated with the Platonic forms, from Plato's theory of forms, according to which the physical objects of sense perception are but mere copies, shadows, of a greater truth (form). Hence, having quantum probability in mind, QL-atoms are described by both an element of chance but also of necessity (as these ideas were discussed in the section titled "The Cycles of the World" in chapter15). But so are the D-atoms; their motion is purely random and chancy, although at the same time what drives them to combine or separate is necessity.

Are the QL-atoms the smallest cuts of matter and, within this context, thus the ultimate D-atoms? It is generally not thought so. Are the QL-atoms different forms of the same type of universal substance, the same type of particle, and what that might be? While the Higgs boson particles have some qualities required of a universal substance, the standard model that predicts them does not include the most puzzling force in the universe, namely, gravity. Therefore, although useful, any model of nature that does not incorporate gravity is incomplete.

The basic concept of the ancient atomic theory was highly valued by Nobel laureate Richard Feynman. He said: "If, in some cataclysm, all scientific knowledge were to be destroyed, and only one sentence passed on to the next generations of creatures, what statement would contain the most [scientific] information in the fewest words? I believe it is the *atomic hypothesis* (or the atomic *fact . . .*) that *all things are made of atoms—little particles that move around in perpetual motion, attracting each other when they are a little distance apart, but repelling upon being squeezed into one another*. In that one sentence, you will see, there is an *enormous* amount of information about the world, if just a little imagination and thinking are applied."[8] On a related note, in his book *The God Particle* Nobel laureate Leon Lederman graded thousands of scientists (including himself) for their efforts in their quest for a primary substance of the universe. He started with Thales all the way to 1993, the completion date of his book. Democritus received the only A in the class![9]

So as a general *idea*, the D-atoms, the uncuttable discrete and fundamental pieces of matter that everything is made of, are still part of our most advanced theories of nature, for these basic but important properties are properties of the QL-atoms, too—in fact also of the strings of string theory. But whether the QL-atoms (together with the force-carrying particles of the standard model as well as the Higgs boson, a total of sixty-one particles all confirmed to exist), a zoo of other unconfirmed particles (including the graviton, predicted by various other scientific models), or some new particles of *one and the same type* (a much-desired scientific simplicity of Democritean grandeur) that manifest themselves by way of the familiar particles are/will be the truly uncuttable discrete and fundamental pieces of matter, remains to be seen. How about the void? Does it exist or not? Is it needed, or can it be avoided?

VOID OR NOT?

The atomists Leucippus and Democritus called an atom a *thing*, Being (what-is), and the void *nothing* (*not thing*), Not-Being (what-is-not).[10] And they agreed with the theory of Parmenides (with one interpretation of it, anyway, for which the properties of Being are understood literally) that motion is impossible without the void. But whereas Parmenides denied the existence of the void by considering it Not-Being, the atomists postulated the opposite: Not-Being, the void, exists, for only then, they thought, can motion and change be accounted for. It is the place to put the atoms and enable them to move. For the atomists the void is empty space, and so *in* it there is nothing. But for Parmenides *it, the void, empty space itself*, is nothing, Not-Being; not *in* it there is nothing. The nature of the void has created mind-boggling debates since the time of Parmenides. For if something, for example, the void, is really nothing, how can it exist? How does one define "nothingness"? The answer is not easy.

But first let's summarize Democritus's arguments favoring the void. By accepting the phenomena of motion, change, and diversity to be real, he deduced the void to be real as well, for without the void his impenetrable, indivisible atoms could not move and consequently the phenomena of change and diversity would not occur; but they do occur, so the void must be real. Similarly, by accepting also multiplicity and division of composite objects to be real, he again deduced void to be real, for without it, composite objects could not be divided (cut) into smaller pieces: "division resulted from the presence of void in bodies."[11] As explained further by philosopher and mathematician Bertrand Russell (by paraphrasing Democritus), "When you use a knife to cut an apple, the knife has to find empty places where it can penetrate; if the apple contained no void, it would be infinitely hard and therefore physically indivisible."[12] Here we recall of course that for Democritus divisibility does not continue ad infinitum; it applies only to composite objects and stops at his physically indivisible atoms. So for Democritus both the atoms and the void are real: "thing [atoms] is [exist] no more than not-thing [void]."[13]

Now what does modern physics think of the void? Does it exist or not? Is it a true nothing, the Parmenidean Not-Being, or something else? While "nature abhors a vacuum,"[14] a popular phrase since the Renaissance, yet "nothing works without, well, nothing."[15]

Void?

On the one hand, void is still a useful concept for the understanding of many phenomena. According to quantum theory, electrons in a chemical atom, for example, "move" around their nucleus by keeping their distance from one another as if space between them is empty, devoid of matter—"a regulation against overcrowding"[16] formally known as the Pauli exclusion principle. As a consequence of this principle, the electrons of chemical atoms keep their distance from each other; they do not like to be squeezed together into a small region, so they act as if they were rigid: the closer they get, the faster they move apart—a statement in agreement with the Heisenberg uncertainty principle, for, according to it, the uncertainties in the position and velocity are inversely proportional, so the smaller a particle's region of confinement (the smaller its position uncertainty), the faster its motion to escape such region (the greater its velocity uncertainty), "almost as if it [the particle] were overcome with claustrophobia," Brian Greene wrote.[17] The exclusion principle explains why chemical atoms are mostly empty space and why macroscopic objects (which are made of chemical atoms) have a degree of rigidity, size, and shape. D-atoms are rigid, consequently, in a sense they, too, obey the regulation against overcrowding, for one D-atom cannot occupy the same region of space as another D-atom. Had the exclusion principle not been true, the QL-atoms, which obey it, would not endure as disconnected pieces of matter, thus nuclei would not form, nor would chemical atoms or the molecules of organic chemistry, and consequently nor would the matter that living things are made of; generally, *all* matter in such a scenario would collapse into a uniform, undifferentiated, and lifeless state. The diversity in nature is in a sense a consequence of the exclusion principle: diversity is a law of nature!

Or Not?

On the other hand (that is, to be able to explain other phenomena), in the quantum realm the void is not really devoid of matter but a very busy place, seething with all-pervasive fields of energy (e.g., light and gravity waves, even the much-required Higgs boson field that explains mass—see section "Worlds without Forces" below), known as vacuum energy. These fields cannot be zero, even in seemingly empty space, because the time-energy uncertainty principle would be in violation (recall the section titled "*Nothing* Comes from Nothing" in chapter 13). And they are actually fluctuating constantly, creating and annihilating pairs of particles with their corresponding antiparticles. These particles, which are called virtual, are not created out of nothing or annihilated into nothing but are made out of energy and return to be energy (the vacuum energy). Unlike real particles, which can be directly observed, virtual particles cannot, even though they can still cause measurable effects on real particles, a proof that "empty" space is really not empty.

Moreover, according to the theory of general relativity, the whole of space is filled by a gravitational field with properties (such as strength) that vary from place to place and from one moment to the next. Einstein explained gravity by assigning *properties* to "empty" space; empty space (and time) is a flexible medium that gets distorted by a mass, and gravity is space's (and time's) distortions. These *properties* are the void, and so in the theory of general relativity the void is not the Parmenidean Not-Being, for Not-Being, a true nothing, is property-less.

Furthermore, an astronomical observation completed in 1998 with the aid of the Hubble Space Telescope found that the expansion of the universe is accelerating—so a galaxy's recession speed measured today is faster than its speed measured yesterday. This accelerated expansion is attributed (although reluctantly) to the existence of dark energy, which is hypothesized to permeate the universe and act as a kind of antigravity by stretching space and causing it to expand at continuously faster speeds. Dark energy, which has not yet been detected, is one of the most puzzling mysteries of the universe. Dark matter is yet another great puzzle: though invisible, for it does not emit light, its existence is inferred indirectly by its gravitational pull on neighboring stars. What makes dark matter invisible, no one knows.

Ordinary matter, matter we can see, which makes up flowers, people, planets, stars, and galaxies, is only about 5 percent of the total stuff in the universe. The other 95 percent, which includes dark energy and dark matter, is stuff that we neither see nor know much about, although their subtle presence is deduced in some way.[18] Space, even "empty" space, is a place of constant, frantic activity of virtual particles, light, gravity, dark energy, dark matter, ordinary matter, of Higgs boson particles, possibly strings, membranes, and who knows what else. It is certainly not the Parmenidean Not-Being (the *absolute* nothing). Therefore, once more we emphasize that no scientific theory can base its beginning, its first cause/s, on absolute nothing, on Not-Being; science must begin from something-ness, and what that might be becomes increasingly more complex. In fact even Leucippus's and Democritus's nothing (their notion of void) is really not nothing, since from the point of view of modern physics "it [their void] was the carrier for geometry and kinematics, making possible the various arrangements and movements of atoms."[19]

Fascinatingly these atomic arrangements and movements were imagined by Democritus to be carried out without the requirement of a force of, say, gravity, electricity, or magnetism; other than their direct collisions during which there was a physical contact, D-atoms experience no other force! D-atoms have no weight, they produce no force of gravity.[20] How can this be? How can there be a world without gravity, without forces in general?

WORLDS WITHOUT GRAVITY

Weight or gravity (that is, the tendency of objects to fall or the property of heaviness) was not one of the primary characteristics of atoms but a property that was accounted for by Democritus ingeniously through motion, in particular rotational motion.[21]

Though motion is chaotic, Democritus argued, in an infinite space with infinite atoms there is always a chance that the bulk of the atoms of a certain region move collectively in a preferred direction of motion, rotational in particular, and produce a vortex. The rotational motion of such a vortex, Democritus thought, ultimately causes the bigger atoms (the more massive, the heavier, as we would say today having gravity in mind) to move toward its center, ultimately forming the earth and the water on it, and the smaller atoms (the lighter) to move toward its outskirts, ultimately forming the air, the sky, and the stars. Because the dynamics of our world system is still rotational (e.g., the sky rotates, relative to us, and so in a sense do the moving clouds), objects made of the bigger atoms, like a rock, still fall, and objects made of the smaller atoms, like steam, smoke, or fire, still rise, Democritus argued. Air, on the other hand, generally does not fall, he thought, because of its rapid revolution, just as water does not spill from a cup when it is rapidly spun around. His analysis was logical as regards observation because the earth, which (for him) is made of the bigger atoms, formed in the center of his vortex, water, made of smaller atoms, is on earth, and air, made of even smaller atoms, is above water and earth.

Now, concerning the dynamics of a vortex, in reality it is the reverse that happens: massive objects tend toward the outskirts of a vortex, and lighter ones toward the center (this, for example, happens in a centrifuge, a device employed to separate different substances). Nonetheless this error in Democritus's explanation is really a minor point compared to the fact that he managed a reasonably clever explanation of the world only in terms of a basic property that atoms have, namely, their motion in the void. Thus he saw no need of a force of a weight, of gravity, despite that apples fall as if a force is pulling them through space. The latter, legend says, inspired Newton to conceive his theory of universal gravitation for which gravity *was* a force, only to be abolished as a force by Einstein's theory of general relativity. How so, and what does quantum theory say about forces in general?

WORLDS WITHOUT FORCES

An apple and the earth, or the sun and the earth, feel a force of attraction from each other, Newton thought, as a consequence of a mysterious action at a distance (not to be confused with the action at a distance of quantum entanglement) that he himself admitted he did not understand. How is gravity transmitted if the interacting objects do not touch each other? How does one body feel the other, how do they communicate, if nothing but empty space exists between them and if nothing specific is really exchanged by them?

Einstein provided the answer through his theory of general relativity. He eliminated the need for an action-at-a-distance-type force by recognizing that the agent that transmits gravity is space itself (in fact, time, too, but let's keep things simple) when distorted by matter. Space is no longer the Newtonian passive playground where events unfold but a flexible medium the geometrical shape of which *changes* (gets warped) by matter. As discussed in chapter 12, the distortion of space (the changing geometry of space) in turn influences an object's motion and feels like gravity. With such geometrical representation of spacetime the notion that gravity is a force is abandoned in general relativity. And in the study of the phenomena of gravity, an object's motion through space and time may no longer be regarded as a response to an action-at-a-distance force of gravity acting on it, as in Newtonian physics, but as a response to the warping spacetime caused by the distribution of all other objects around it.

Moreover, as already discussed in chapter 12, according to the standard model of quantum theory, the particles of matter, the QL-atoms, combine with one another via the continual exchange of the particles of force. Recall, for example, that the attractive and repulsive electric force is really a manifestation of intricate particle collisions; even gravity is hypothesized to work via the exchange of gravitons. Matter and force are no longer distinct notions. Instead, forces are really expressions of complicated particle collisions.

And so, as is the view of Democritus, nature can be understood in terms of just particles and their complex collisions—forces were never required in the theory of Democritus and are no longer required in modern physics! Incidentally, although Empedocles's two forces, love and strife, were separate entities from his four elements, still they were not action-at-a-distance-type of force:

via their direct contact, they either pushed the elements together to mix or pushed them apart to separate, so they, too, in a way, acted as colliding particles.

Even mass, and consequently weight, is thought to not be a fundamental property of the QL-atoms, rather, a property the QL-atoms acquire through their interactions with the Higgs boson field. The mass of an object is a measure of its resistance to motion. The smaller the resistance, the less the mass is. Throwing a baseball is easier than a bowling ball: the baseball has less mass than a bowling ball, or, equivalently, it produces less resistance to our attempt to throw it. Now, the Higgs boson field pulls on the other particles (e.g., the QL-atoms) as they traverse through it and impedes their motion. It is this resistance that we interpret as mass. In an analogy, stirring a cup of coffee with a spoon is easy, but a cup of honey is not. Honey is a more viscous fluid, and the spoon feels heavier, more massive, as it moves through it. In this analogy, the spoon is a QL-atom and the fluid the Higgs boson field. Just as fluids with different viscosities cause the spoon that moves through them to feel lighter or heavier by impeding its motion, the standard model imagines that the all-pervasive Higgs boson field creates an analogous effect on the initially massless QL-atoms as they traverse it, endowing each with a unique mass and slowing them down. It is as if the Higgs field had different viscosity for different-type QL-atoms. Similarly, the force-carrying particles W's and Z's acquire their mass, but photons, feeling no resistance by the Higgs field, remain massless and thus can move with the speed of light. The analogy describes what is known as the Higgs mechanism, which explains why some particles have mass and some not (though it does not explain why they have the actual mass they do). What particular agent gives Higgs bosons their mass is nevertheless still an unknown. The idea that mass may not be a fundamental property was prompted by a few interesting and unresolved questions. Why, for example, is there no pattern in the mass values of the particles, a fact in contrast to other particle properties, such as spin or electric charge? For instance, in some units of measurement, the spin of all QL-atoms is $1/2$ and the spin of all the force-carrying particles is 1.

Amazingly, mass is not a fundamental particle property, neither in the standard model of modern physics nor in Democritus's atomic theory! Equally amazing is that in both the modern and the Democritean physics, the non-fundamental property of mass (and the consequent heaviness, weight) is caused (is acquired) by atomic motion—the motion of the QL-atoms through the Higgs field, and the motion of the D-atoms in the vortex!

Particles are by definition discrete entities, thus their existence implies a certain discontinuity in nature. But is the nature of nature truly discontinuous?

CONTINUITY VERSUS DISCONTINUITY

If indeed something does exist always everywhere (including apparently empty space), then the essence of existence is continuous. At the same time, to make sense of the diversity in nature, the continuity of that which exists must vary, from place to place and from time to time. These variations are interpreted as discontinuities in matter and energy and are called particles. But what separates these discontinuities cannot be absolute nothingness, for energy is ever-present and everywhere. If the sea is the energy, the sea waves are the fluctuations of the energy, that is, the discrete particles of matter and the discrete particles of force. But even between the sea waves there exists water, the sea, energy, not nothing. So the view of modern physics is some kind of combination of the Parmenidean Being (of an indivisible, continuous whole obeying one eternal truth), the Heraclitean constant change (of everything in the sensible world), and the Democritean discreteness (of a whole, which while in essence continuous is also inhomogeneous and discrete, for it fluctuates). Now what exists must, we believe, be describable by a single idea or equation, a single type of particle. Can the human intellect ever conceive it? What is the role of the senses in conceiving it?

INTELLECT VERSUS SENSES

For Democritus reality is objective and much deeper than what's revealed by sense perception alone. Trying to capture both the unreliability and the significance of sense perception in our attempts to understand nature rationally, Democritus imagined a hypothetical dialogue between the intellect and the senses.

Intellect: "Sweet is by convention, bitter by convention, hot by convention, cold by convention, color by convention, in reality however there are but atoms and the void."[22]

Senses: "Troubled Intellect! From us you take the evidence and you want to overthrow us? Our fall will be your fall."[23]

The *Intellect* says that what's perceived by the *Senses* is radically different from the way nature really is. Knowledge derived by the *Senses* is "bastard" but by the *Intellect* "legitimate"[24] (Democritus). The *Senses* perceive sight, hearing, smell, taste, and touch, but these sensations are not objective properties of nature. They are only perceptions by convention (in relation to us), only a consequence of the atoms and their motion in the void—the objective truth, in other words, the true nature of things, is only atoms and the void the *Intellect* claims.

This might be true, the *Senses* respond, but though unreliable ("bastard"), the quest for knowledge always begins with sense perception. For the evidence of the atoms and the void is obtained through observation of colors, tastes, and so on, thus the participation of the *Senses*. It is what we see that we use in order to explore what we cannot see, the *Senses* emphasize. After all, "the phenomena [what is seen, occurrences] are a sight of the unseen."[25] At the end, neither the intellect alone nor the senses alone can lead to the truth, but their combination might.

Remarkably, for Democritus the aggregations and segregations of unseen atoms in the void produced not only our own world (with the earth, moon, sun, planets, stars, plants, fish, animals, and including humans) but also count-less others.

WORLDS OTHERWORLDLY

The process that creates a vortex from which a world like ours was formed is not unique in the universe, Democritus thought, so he posited the existence of otherworldly worlds. (Democritus therefore, like the Pythagoreans and pos-sibly Anaxagoras, was not fixated on a geocentric model.) In modern termi-nology such worlds are really the galaxies; the stars we see at night all belong to the Milky Way galaxy, which is one of numerous others in the universe. The idea of multiple worlds does not violate Democritus's isotropic universe, where there is no preferred direction of motion, hence all directions are equally probable. The clockwise rotation of a vortex that creates one world is canceled out by the counterclockwise rotation of a neighboring world, a phenomenon that modern physics calls the conservation of angular momentum.

Today we know that our universe consists of many galaxies, each of which contains billions of stars, each of which may have its own group of planets. In fact earthlike (thus habitable) planets are common! According to the latest research one in five sun-like stars in our Milky Way galaxy has an earthlike planet orbiting it—the nearest case is perhaps of a star only twelve light-years away and visible to the naked eye.[26] Consequently the existence of life elsewhere in the universe besides earth is increasingly more probable. While scientists have not yet discovered any, still one may ponder such a question optimistically. First, one may answer favorably on the grounds of modesty. Our part of the universe is no more special than another (the laws of nature are everywhere the same), hence it is not unreasonable to expect life to evolve elsewhere as it did on earth. This is especially so on earthlike extrasolar planets (planets orbiting other stars). Or, one may answer mathematically by using the famous Drake equation to calculate the odds for intelligent life elsewhere in the universe.[27]

The equation gives a rough estimate of N, the number of advanced civilizations present in the Milky Way galaxy with which we might be able to establish radio communication. The important factors that N depends on are:

$N_{\text{-habitable planets}}$ = the number of habitable planets in the galaxy—those capable of developing and sustaining life;

$f_{\text{-life}}$ = the fraction of habitable planets where life actually evolves;

$f_{\text{-civilization}}$ = the fraction of planets with life where also an advanced civilization (with capabilities of interstellar communication) has developed at some point in the past;

$f_{\text{-now}}$ = the fraction of planets that still have an advanced civilization, now.

N is equal to the product of the above factors, that is:

$$N = N_{\text{-habitable planets}} \times f_{\text{-life}} \times f_{\text{-civilization}} \times f_{\text{-now}}$$

None of these numbers is known with certainty. We can only make rough estimates. Suppose, in a sample exercise, that there exist 1,000 habitable planets in the Milky Way, in other words, $N_{\text{-habitable planets}}$ = 1,000. And life evolves on 1 in 10 of them, in other words, $f_{\text{-life}}$ = 1 / 10 = 0.1. Now, 1 in 4 of these planets with life, also develops, sometime, an advanced civilization, in other words, $f_{\text{-civilization}}$ = 1 / 4 = 0.25. Finally only 1 in 5 of such advanced civilizations are still around today, in other words, $f_{\text{-now}}$ = 1 / 5 = 0.2. What is N?

N = 1,000 × 0.1 × 0.25 × 0.2 = 5

How do you think the human race will react to a discovery of an advanced extraterrestrial civilization? What would it be like to be exposed to another intelligent species' worldview? We have no experience with that since on earth, out of the myriad species, only humans have developed advanced thought. Why?

CONCLUSION

In an attempt to understand nature in a logical and causal manner, the search for the primary substance of the universe in pre-Socratic philosophy comes to an end with the atomic theory of Leucippus and Democritus. Their notion of indivisible, discrete particles without substructure has endured and, according to modern physics, is still one of the most remarkable properties of nature. Could spacetime form a type of discreteness, too? Namely, is there a fundamental irreducible (smallest, without substructure, and finitely divisible) space length and time interval, or is there always a smaller portion of them (i.e., are they infinitely divisible)? Although no experiment has confirmed spacetime discreteness, some modern theoretical models speculate that it might be true. Recall Democritus's notion of atoms was confirmed two and a half millennia after it was first proposed. Interestingly, the hypothesis of space discreteness has been an innovation of Epicurus (341–270 BCE), a post-Socratic, who continued the remarkable work of Democritus.

EPILOGUE

Our ancient quest for knowledge began with our evolution two hundred thousand years ago, and with everything we experienced through our struggles to survive and our efforts to thrive and live fully. We hunted and gathered, painted on caves, told stories, domesticated animals and plants, built homes, wondered about nature, gave birth to civilization and religion, picked up writing, philosophized, and engaged in science. In fact, we keep on doing all these wonderful things, but, amazingly, we have been doing them ever more in the light of science.

Since its birth 2,600 years ago, science has evolved significantly. Nonetheless its ultimate goal still remains essentially the same: to understand nature rationally and to reduce the explanations of all natural phenomena to the least possible number of basic assumptions (first causes, axioms)—ideally to just one. Now, say that has been achieved, will the human intellect be satisfied?

We like the Homer's *Odyssey* so much because it is a story of a journey (in fact a long one) not of a destination. Shortly after Odysseus returns to Ithaca the story ends and we all get melancholy—we like the journey better than the destination, for although, with Odysseus's return, the events in Ithaca were breathtakingly exciting and awaited eagerly from the start of the story, their completion also brought the absolute end of the epic adventure.

The beauty of nature is in her secrets, the magic is in our discoveries. I never want to know everything—to have the journey of knowledge, of search and discovery, ever end. What would be next if I did? What would happen to the magic? Space, time, matter, energy, the human senses to observe, and the intellect to contemplate—it is all nature, and her nature is her many secrets (her shadows). They are many but also intelligible (steal-able, like Promethean fire)! I hope you have a magical, endless journey searching, in the light of science, for the nature of nature!

NOTES

PROLOGUE

1. The Higgs boson, nicknamed "God Particle" by Leon Lederman for its elusiveness as well as significance for our understanding of the structure of matter. Leon Lederman and Dick Teresi, *The God Particle: If the Universe Is the Answer, What Is the Question?* (Boston: Houghton Mifflin, 1993).

CHAPTER 2. WHAT IS SCIENCE?

1. Dionysius, bishop of Alexandria, from Eusebius, *Preparation for the Gospel* 14.7.4, trans. Daniel W. Graham, *The Texts of Early Greek Philosophy: The Complete Fragments and Selected Testimonies of the Major Presocratics* (Cambridge: Cambridge University Press, 2010), p. 521 (text 5). Graham's sourcebook contains fragments and testimonies of the major pre-Socratic philosophers. It is a Greek-English edition that also cites the original ancient source of each text.

CHAPTER 3. URBANIZATION

1. Pliny *Natural History* 2.53. See Daniel W. Graham, *The Texts of Early Greek Philosophy: The Complete Fragments and Selected Testimonies of the Major Presocratics* (Cambridge: Cambridge University Press, 2010), p. 25 (text 5).

2. Smithsonian National Museum of Natural History, http://humanorigins.si.edu/ (accessed March 10, 2014).

3. Peter Tyson, "Who's Who In Human Evolution," http://www.pbs.org/wgbh/nova/evolution/whos-who-human-evolution.html (accessed March 10, 2014).

4. John Savino and Marie D. Jones, *Supervolcano: The Catastrophic Event That Changed the Course of Human History* (NJ: Career Press, 2008); Smithsonian National Museum of Natural History, http://humanorigins.si.edu/evidence/human-evolution-timeline-interactive (accessed June 6, 2014); Toba: The Toba Super-Eruption, http://toba.arch .ox.ac.uk/index.htm (accessed June 6, 2014).

5. Ian Tattersall, *The World from Beginnings to 4000 BCE* (Oxford: Oxford University Press, 2008), p. 89; Savino and Jones, *Supervolcano*, p. 126; Smithsonian National Museum

of Natural History, http://humanorigins.si.edu/human-characteristics/change (accessed March 10, 2014).

6. Richard Wrangham, *Catching Fire: How Cooking Made Us Human* (New York: Basic Books, 2009).

7. Sister Wendy Beckett, *The Story of Painting* (New York: Dorling Kindersley, 2000), p. 10.

8. Ibid., p. 11.

9. Tattersall, *World from Beginnings*, p. 107; Tyson, "Who's Who In Human Evolution," http://www.pbs.org/wgbh/nova/evolution/whos-who-human-evolution.html (accessed June 6, 2014); Smithsonian National Museum of Natural History, http://humanorigins. si.edu/evidence/human-fossils/species/homo-floresiensis (accessed June 6, 2014).

10. Walter Burkert, *Greek Religion: Archaic and Classical*, trans. John Raffan (Cambridge, MA: Blackwell Publishing Ltd. and Harvard University Press, 1985), p. 47.

11. Ibid., p. 248.

12. Isaac Asimov, *Asimov's Chronology of Science and Discovery* (New York: HarperCollins, 1989), p. 12.

13. Hesiod *Works and Days* 120, trans. Hugh G. Evelyn, *Mythology Ultimate Collection* (Houston: Everlasting Flames Publishing, 2010).

14. Ibid., pp. 127–28.

15. Ibid., pp. 133–37.

16. Gen. 4:1–26.

17. Hesiod *Works and Days* 110–20.

18. Ibid., 127–28.

CHAPTER 4. THE MYTHOLOGICAL ERA

1. Bertrand Russell, *The History of Western Philosophy* (New York: Simon & Schuster, 1945), p. 11.

2. Plato *Republic* 3.390E, trans. Allan Menzies, *History of Religion: A Sketch of Primitive Religious Beliefs and Practices, and of the Origin and Character of the Great Systems* (Memphis, TN: General Books, 2010), p. 40.

3. Hesiod *Theogony* 542, trans. Hugh G. Evelyn, *Mythology Ultimate Collection* (Houston: Everlasting Flames Publishing, 2010).

4. Hesiod *The Astronomy* fragment 4. See Evelyn, *Mythology Ultimate Collection*.

5. Apollodorus (pseudo-Apollodorus) Book 1 of Bibliotheca. See Darryl Marks, *Mythology Ultimate Collection* (Houston: Everlasting Flames Publishing, 2010).

6. Gen. 1:27.

7. Apollodorus (pseudo-Apollodorus) Book 1 of Bibliotheca.

8. Sextus Empiricus *Against the Professors* 9.19. See Daniel W. Graham, *The Texts of Early Greek Philosophy: The Complete Fragments and Selected Testimonies of the Major Presocratics* (Cambridge: Cambridge University Press, 2010), p. 613 (text 187).

9. Gen. 14–15.

10. Quoted in Erwin Schrödinger, *Nature and the Greeks and Science and Humanism* (Cambridge: Cambridge University Press, 1996), p. 69.

11. Kathleen Freeman, *Ancilla to the Pre-Socratic Philosophers* (Cambridge, MA: Harvard University Press, 1996), p. 23.

12. Ibid., p. 22.

13. Clement of Alexandria 7.22, trans. Graham, *Texts of Early Greek Philosophy*, p. 109 (texts 31, 32), p. 111 (text 33).

14. Martin West, "Early Greek Philosophy," in *The Oxford Illustrated History of Greece and the Hellenistic World* 1986, ed. John Boardman, Jasper Griffin, and Oswyn Murray (Oxford: Oxford University Press, 1986), p. 110.

CHAPTER 5. RELIGION AND SCIENCE

1. Karl R. Popper, *Conjectures and Refutations: The Growth of Scientific Knowledge* (London and New York: Routledge, 1989), p. 50.

2. Robert Parker, "Greek Religion," in *The Oxford Illustrated History of Greece and the Hellenistic World 1986*, ed. John Boardman, Jasper Griffin, and Oswyn Murray (Oxford: Oxford University Press, 1986), p. 248.

3. Walter Burkert, *Greek Religion: Archaic and Classical*, trans. John Raffan (MA: Blackwell Publishing Ltd. and Harvard University Press, 1985), pp. 276–304.

4. Bertrand Russell, *The History of Western Philosophy* (New York: Simon & Schuster, 1945), p. xiii.

CHAPTER 6. THE BIRTH OF SCIENCE

1. Andrew Gregory, *Eureka! The Birth of Science* (Cambridge: Icon Books, 2001); Bertrand Russell, *The History of Western Philosophy* (New York: Simon & Schuster, 1945); Carl Sagan, *Cosmos* (New York: Random House, 1980), Erwin Schrödinger, *Nature and the Greeks and Science and Humanism* (Cambridge: Cambridge University Press, 1996); G. E. R. Lloyd, *Early Greek Science: Thales to Aristotle* (New York: W. W. Norton & Company, 1970); G. S. Kirk, J. E. Raven, and M. Schofield, *The Presocratic Philosophers* (Cambridge: Cambridge University Press, 1983); Isaac Asimov, *The Greeks; A Great Adventure* (Boston: Houghton Mifflin, 1965); John Burnet, *Early Greek Philosophy* (London: A & C Black, 1920).

2. Russell, *History of Western Philosophy*, p. 10.

3. Smithsonian National Museum of Natural History, http://humanorigins.si.edu/ evidence/human-family-tree (accessed March 10, 2014); Peter Tyson, "Who's Who In Human Evolution," http://www.pbs.org/wgbh/nova/evolution/whos-who-human -evolution.html (accessed March 10, 2014).

4. Smithsonian National Museum of Natural History, http://humanorigins.si.edu/evidence/human-fossils/species (accessed March 10, 2014).

5. Jared Diamond, *The Third Chimpanzee* (New York: HarperCollins, 1992); Jennie Cohen, "Did Neanderthals Create World's Oldest Cave Painting?" http://www.history.com/news/did-neanderthals-create-worlds-oldest-cave-paintings (accessed June 7, 2014).

6. Discovery Channel, "Neanderthal," http://www.youtube.com/watch?v=W7UFbxsF3p0&feature=related (accessed June 7, 2014); History Channel, "Clash of the Cavemen DVD," http://www.youtube.com/watch?v=gUifwntZBZw&feature=related (accessed June 7, 2014).

7. Diamond, *Third Chimpanzee*; Discovery Channel, "Neanderthal"; History Channel, "Clash of the Cavemen DVD."

8. Ibid.

9. Isaac Asimov, *Asimov's Chronology of the World* (New York: HarperCollins, 1991).

10. Daniel R. Altschuler and Christopher J. Salter, "The Arecibo Observatory: Fifty Astronomical Years," *Physics Today* 66, no. 11 (November 2013): 45; Sagan, *Cosmos*, p. 290.

11. Oswyn Murray, "Life and Society in Classical Greece," in *The Oxford Illustrated History of Greece and the Hellenistic World 1986*, ed. John Boardman, Jasper Griffin, and Oswyn Murray (Oxford: Oxford University Press, 1986), p. 221.

12. Walter Burkert, *Greek Religion: Archaic and Classical*, trans. John Raffan (MA: Blackwell Publishing Ltd. and Harvard University Press, 1985), p. 4.

13. Russell, *History of Western Philosophy*, p. 208.

14. Schrödinger, *Nature and the Greeks*, p. 84.

15. Erwin Schrödinger, *What Is Life? & Mind and Matter* (Cambridge: Cambridge University Press, 1967).

16. Plato *Phaedrus* 246A–254E. See trans. Benjamin Jowett, *The Complete Works of Plato* (The Complete Works Collection, 2011).

17. Russell, *History of Western Philosophy*, p. 16.

18. Plato *Phaedrus* 245A, trans. Jowett, *The Complete Works of Plato*, Kindle Locations 16329–30.

19. Russell, *History of Western Philosophy*, p. 21.

20. Edith Hamilton, *The Greek Way* (New York and London: W. W. Norton & Company, 1930), p. 224.

21. Bruce Thornton, *Greek Ways: How the Greeks Created Western Civilization* (San Francisco: Encounter Books, 2002), p. 4.

22. Stephen Bertman, *The Genesis of Science: The Story of Greek Imagination* (Amherst, NY: Prometheus Books, 2010), Kindle Locations 52–53.

CHAPTER 7. CLOSE ENCOUNTER OF THE TENTH KIND

1. This chapter was inspired by chapter 2 of *The God Particle*, in which Leon Lederman imagines conversing with Democritus: Leon Lederman and Dick Teresi, *The God Particle: If the Universe Is the Answer, What Is the Question?* (Boston: Houghton Mifflin, 1993).

CHAPTER 8. THALES AND SAMENESS

1. Aristotle *Metaphysics* 983b6–13, 17–27. See Daniel W. Graham, *The Texts of Early Greek Philosophy: The Complete Fragments and Selected Testimonies of the Major Presocratics* (Cambridge: Cambridge University Press, 2010), p. 29 (text 15); Aëtius 1.31, 1.10.12. See Graham, *Texts of Early Greek Philosophy*, p. 29 (text 16); Simplicius *Physics* 23.21–29. See Graham, *Texts of Early Greek Philosophy*, p. 29 (text 17).

2. Aristotle *Metaphysics* 983b6–13, 17–27. See Graham, *Texts of Early Greek Philosophy*, p. 29 (text 15).

3. Aëtius 1.31, 1.10.12. See Graham, *Texts of Early Greek Philosophy*, p. 29 (text 16).

4. Ibid.

5. Ibid; Aristotle *Metaphysics* 983b6–13, 17–27. See Graham, *Texts of Early Greek Philosophy*, p. 29 (text 15).

6. Stephen Hawking, *A Brief History of Time: From the Big Bang to Black Holes* (New York: Bantam Books, 1988), chap. 5.

7. Brian Greene, *The Elegant Universe: Superstrings, Hidden Dimensions, and the Quest for the Ultimate Theory* (New York: W. W. Norton & Company, 1999), p. 144.

8. Gareth Morgan, "Stephen Hawking Says There Is No Such Thing as Black Holes, Einstein Spinning in His Grave," Express, January 24, 2014, http://www.express.co.uk/news/science-technology/455880/Stephen-Hawking-says-there-is-no-such-thing-as-black-holes-Einstein-spinning-in-his-grave (accessed March 12, 2014); Zeeya Merali, "Stephen Hawking: 'There Are No Black-Holes,'" *Nature*, January 24, 2014, http://www.nature.com/news/stephen-hawking-there-are-no-black-holes-1.14583 (accessed March 12, 2014).

9. Aristotle *On the Soul* 411a7–8, trans. Graham, *Texts of Early Greek Philosophy*, p. 35 (text 35).

10. Diogenes Laertius 1.24. See Graham, *Texts of Early Greek Philosophy*, p. 21 (text 1).

11. Plato *Theaetetus* 174a4–8, trans. Graham, *Texts of Early Greek Philosophy*, p. 25 (text 7).

12. Dante *Divine Comedy*, trans. BookCaps (BookCaps Study Guides, 2013), Kindle locations 10900–901.

13. Aristotle *Politics* 1259a5–21, trans. Graham, *Texts of Early Greek Philosophy*, p. 25 (text 8).

14. Diodorus of Sicily 1.39.1–3. See Graham, *Texts of Early Greek Philosophy*, p. 563 (text 84).

15. Aristotle *Meteorology* 342b25, trans. Demetris Nicolaides. See also Graham, *Texts of Early Greek Philosophy*, p. 303 (text 48).

CHAPTER 9. ANAXIMANDER AND THE INFINITE

1. Leon Lederman and Dick Teresi, *The God Particle: If the Universe Is the Answer, What Is the Question?* (Boston: Houghton Mifflin, 1993).

2. Werner Heisenberg, *Physics and Philosophy: The Revolution in Modern Science* (New York: Harper Torchbooks, 1962), p. 36.

3. Lederman and Teresi, *God Particle*, p. 56.

4. In the Kelvin scale the absolute zero is 0 degrees Kelvin, which is -273.15 degrees Celsius, which is -459.67 Fahrenheit.

5. Aristotle *On the Heavens* 295b10–16. See Daniel W. Graham, *The Texts of Early Greek Philosophy: The Complete Fragments and Selected Testimonies of the Major Presocratics* (Cambridge: Cambridge University Press, 2010), p. 59 (text 21).

6. Karl R. Popper, *Conjectures and Refutations: The Growth of Scientific Knowledge* (London and New York: Routledge, 1989) p. 138.

7. John Burnet, *Early Greek Philosophy* (London: A & C Black, 1920), chap. 1.

8. Charles Sherrington, *Man on His Nature* (Cambridge: Cambridge University Press, 2009), p. 302.

9. Erwin Schrödinger, *Nature and the Greeks and Science and Humanism* (Cambridge: Cambridge University Press, 1996), p. 66.

10. Heisenberg, *Physics and Philosophy*, p. 128.

11. Richard P. Feynman, *Six Easy Pieces* (New York: Perseus Publishing, 1963), p. 22.

12. Aëtius 5.19.4. See Graham, *Texts of Early Greek Philosophy*, p. 63 (text 37); Censorinus 4.7. See Graham, *Texts of Early Greek Philosophy*, p. 63 (text 38); Hippolytus *Refutation* 1.6.6. See G. S. Kirk, J. E. Raven, and M. Schofield, *The Presocratic Philosophers* (Cambridge: Cambridge University Press, 1983), Kindle Location 3598; Plutarch *Symposium* 730e. See Graham, *Texts of Early Greek Philosophy*, p. 63 (text 39); Ps.- Plutarch *Strom.* 2. See Kirk, *Presocratic Philosophers*, Kindle Location 3590.

13. Censorinus 4.7; Hippolytus *Refutation* 1.6.6; Plutarch *Symposium* 730e; Ps.- Plutarch *Strom.* 2.

14. Aëtius 5.19.4; Hippolytus *Refutation* 1.6.6.

CHAPTER 10. ANAXIMENES AND DENSITY

1. Simplicius *Physics* 24.26–25.1, Theophrastus frag. 226A. See Daniel W. Graham, *The Texts of Early Greek Philosophy: The Complete Fragments and Selected Testimonies of the Major Presocratics* (Cambridge: Cambridge University Press, 2010), p. 75 (text 3).

2. Aëtius 1.3.4, trans. John Burnet, *Early Greek Philosophy* (London: A & C. Black, 1920), chap. 1.

3. Hippolytus *Refutation* 1.7, trans. Burnet, *Early Greek Philosophy*, chap. 1.

4. Aëtius 3.3.2, trans. Burnet, *Early Greek Philosophy*, chap. 1.

5. This view is also expressed in Erwin Schrödinger, *Nature and the Greeks and Science and Humanism* (Cambridge: Cambridge University Press, 1996); Werner Heisenberg, *Physics and Philosophy: The Revolution in Modern Science* (New York: Harper Torchbooks, 1962).

6. Sextus Empiricus *Against the Professors* 7.135, trans. Schrödinger, *Nature and the Greeks*, p. 89.

7. Aristotle *Metaphysics* 985b4–20, trans. Graham, *Texts of Early Greek Philosophy*, p. 525 (text 10).

8. Ibid.

9. Carl Sagan, *Cosmos* (New York: Random House, 1980), p. 181.

10. Schrödinger, *Nature and the Greeks*, pp. 62–65, 84–86, 157–62.

11. Ibid., p. 160.

CHAPTER 11. PYTHAGORAS AND NUMBERS

1. Aristotle *Metaphysics* 987b22. See Erwin Schrödinger, *Nature and the Greeks and Science and Humanism* (Cambridge: Cambridge University Press, 1996), p. 35.

2. Diogenes Laertius 8.46. See G. S. Kirk, J. E. Raven, and M. Schofield, *The Presocratic Philosophers* (Cambridge: Cambridge University Press, 1983), Kindle Locations 9294–95.

3. Aëtius 2.1.1, trans. Demetris Nicolaides. See Greek book Βας. Α. Κύρκος, *Οι Προσωκρατικοί: Οι Μαρτυρίες και τα Αποσπάσματα τόμος Α* (Αθήνα: Εκδόσεις Δημ. Ν. Παπαδήμα, 2005), p. 247.

4. Aristotle *On the Heavens* 290b12, trans. Demetris Nicolaides. See also Kirk, Raven, and Schofield, *Presocratic Philosophers*, Kindle Locations 9131–33.

5. Johannes Kepler, *The Harmonies of the World*, quoted in George N. Gibson and Ian D. Johnston, "New Themes and Audiences for the Physics of Music," *Physics Today* 55, no. 1 (January 2002): 44.

6. Arnold Sommerfeld quoted in Gibson and Johnston, "New Themes and Audiences for the Physics of Music," *Physics Today* 55, no. 1 (January 2002): 43.

7. *The Elegant Universe: Part 1*, PBS, October 28, 2003.

8. Schrödinger, *Nature and the Greeks*, p. 122; Werner Heisenberg, *Physics and Philosophy: The Revolution in Modern Science* (New York: Harper Torchbooks, 1962), p. 46.

9. Schrödinger, *Nature and the Greeks*, p. 45.

10. Ibid.

11. Bertrand Russell, *The History of Western Philosophy* (New York: Simon & Schuster, 1945), p. 214.

12. Ibid., p. 217.

13. Ibid., p. 540.

CHAPTER 12. HERACLITUS AND CHANGE

1. Diogenes Laertius 2.22, trans. Daniel W. Graham, *The Texts of Early Greek Philosophy: The Complete Fragments and Selected Testimonies of the Major Presocratics* (Cambridge: Cambridge University Press, 2010), p. 181 (text 163).

2. Origen *Against Celsus* 6.42, trans. Graham, *Texts of Early Greek Philosophy*, p. 157 (text 58).

3. Aristotle *Eudemian Ethics* 1235a25–29, trans. Graham, *Texts of Early Greek Philosophy*, p. 157 (text 60).

4. Hippolytus *Refutation* 9.9.2. See Bertrand Russell, *The History of Western Philosophy* (New York: Simon & Schuster, 1945), p. 43.

5. Themestius *Orations* 5.69b, trans. Demetris Nicolaides. See Graham, *Texts of Early Greek Philosophy*, p. 161 (text 75).

6. Erwin Schrödinger, *Nature and the Greeks and Science and Humanism* (Cambridge: Cambridge University Press, 1996), p. 157.

7. Leon Lederman and Dick Teresi, *The God Particle: If the Universe Is the Answer, What Is the Question?* (Boston: Houghton Mifflin, 1993).

8. Planck's constant is a very small number equal to, 6.63×10^{-34} joules $\times$ seconds.

9. Plato *Cratylus* 402a8–10, trans. Graham, *Texts of Early Greek Philosophy*, p. 159 (text 63).

10. Clement *Miscellanies* 5.103.6, trans. Demetris Nicolaides. See Graham, *Texts of Early Greek Philosophy*, p. 155 (text 47).

11. Ibid., 5.104.3–5, trans. John Burnet, *Early Greek Philosophy* (London: A & C Black, 1920), chap. 3.

12. Plutarch *On the E at Delphi* 338d–e, trans. Graham, *Texts of Early Greek Philosophy*, p. 157 (text 55).

13. Clement *Miscellanies* 5.103.6. See Graham, *Texts of Early Greek Philosophy*, p. 155 (text 47).

14. Karl R. Popper, *Conjectures and Refutations: The Growth of Scientific Knowledge* (London and New York: Routledge, 1989) p. 147.

15. Werner Heisenberg, *Physics and Philosophy: The Revolution in Modern Science* (New York: Harper Torchbooks, 1962), p. 37.

16. Ibid., p. 45.

17. Heraclitus *Homeric Questions* 24, trans. Demetris Nicolaides. See Graham, *Texts of Early Greek Philosophy*, p. 159 (text 65).

18. Schrödinger, *Nature and the Greeks*, pp. 123–25.

CHAPTER 13. PARMENIDES AND ONENESS

1. Clement *Miscellanies* 6.23, trans. Erwin Schrödinger, *Nature and the Greeks and Science and Humanism* (Cambridge: Cambridge University Press, 1996), p. 27.

2. Aristotle *Metaphysics* 985b4–20, trans. Daniel W. Graham, *The Texts of Early Greek Philosophy: The Complete Fragments and Selected Testimonies of the Major Presocratics* (Cambridge: Cambridge University Press, 2010), p. 525 (text 10).

3. Ibid.

4. Einstein quoted in Joanne Baker, *50 Physics Ideas You Really Need to Know* (UK: Quercus, 2007), p. 165.

5. Werner Heisenberg, *Physics and Philosophy: The Revolution in Modern Science* (New York: Harper Torchbooks, 1962).

6. J. J. Sakurai, *Modern Quantum Mechanics* (CA: The Benjamin / Cummings Publishing Company, Inc., 1985), pp. 226–29.

7. Graham, *Texts of Early Greek Philosophy*, pp. 211–19.

8. Ibid., pp. 219–33.

CHAPTER 14. ZENO AND MOTION

1. Aristotle *Physics* 239b9–14, trans. Daniel Kolac and Garrett Thomson, *The Longman Standard History of Philosophy* (New York: Pearson, 2005), p. 33.

2. Elias *Commentary on Aristotle's Categories* 109.20–22. See Richard D. McKirahan, *Philosophy before Socrates* (Indianapolis/Cambridge: Hackett Publishing, 2010), p. 182 (Kindle ed.).

3. Aristotle *Physics* 239b14–20, trans. Demetris Nicolaides. See also Daniel W. Graham, *The Texts of Early Greek Philosophy: The Complete Fragments and Selected Testimonies of the Major Presocratics* (Cambridge: Cambridge University Press, 2010), p. 261 (text 18).

4. Ibid., 239b30–33. See Graham, *Texts of Early Greek Philosophy*, p. 261 (text 19); ibid., 239b5–9. See Graham, *Texts of Early Greek Philosophy*, p. 261 (text 20); Diogenes Laertius 9.72. See Graham, *Texts of Early Greek Philosophy*, p. 261 (text 21).

5. Robert J. Oppenheimer, *Science and the Common Understanding* (New York: Simon & Schuster, 1954), p. 40.

6. Aristotle *Physics* 209a23–25, trans. Demetris Nicolaides. See also Graham, *Texts of Early Greek Philosophy*, p. 263 (text 24).

7. Simplicius *Physics* 140.34–141.8. See Graham, *Texts of Early Greek Philosophy*, p. 255 (text 7); ibid., 140.27–34. See Graham, *Texts of Early Greek Philosophy*, p. 259 (text 13).

CHAPTER 15. EMPEDOCLES AND ELEMENTS

1. Leon Lederman and Dick Teresi, *The God Particle: If the Universe Is the Answer, What Is the Question?* (Boston: Houghton Mifflin, 1993), p. 340.

2. Ibid., p. 173.

3. Simplicius *Physics* 158.1–159.4. See also Daniel W. Graham, *The Texts of Early Greek Philosophy: The Complete Fragments and Selected Testimonies of the Major Presocratics* (Cambridge: Cambridge University Press, 2010), p. 251 (text 41).

4. Jeffrey Bennett, Megan Donahue, Nicholas Schneider, and Mark Void, *The Essential Cosmic Perspective 7th Edition* (Boston: Pearson, 2013), p. 450.

5. Aristotle *On Youth, Old Age, Life, Death, and Respiration* 473b9–474a6. See Graham, *Texts of Early Greek Philosophy*, p. 387 (text 127).

6. Simplicius *On the Heavens* 529.1–17, trans. Graham, *Texts of Early Greek Philosophy*, p. 361 (text 51).

7. Ibid.

8. Ibid., trans. Bertrand Russell, *The History of Western Philosophy* (New York: Simon & Schuster, 1945), p. 54.

9. Ibid., 586.12, 587.1–2, trans. John Burnet, *Early Greek Philosophy* (London: A & C Black, 1920), chap. 7 (frag. 57).

CHAPTER 16. ANAXAGORAS AND NOUS

1. Simplicius *Physics* 164.24–25, 156.13–157.4, 176.34–177.6. See also Daniel W. Graham, *The Texts of Early Greek Philosophy: The Complete Fragments and Selected Testimonies of the Major Presocratics* (Cambridge: Cambridge University Press, 2010), p. 291 (text 31).

2. Ibid., 164.23–24, trans. G. E. R. Lloyd, *Early Greek Science: Thales to Aristotle* (New York: W. W. Norton & Company, 1970), p. 44.

3. Ibid., 27.2–23. See Gregory Vlastos, *Studies in Greek Philosophy. Vol. 1: The Presocratics* (Princeton: Princeton University Press , 1993), p. 319.

4. John Burnet, *Early Greek Philosophy* (London: A & C Black, 1920), chap. 6; Vlastos, *Studies in Greek Philosophy*, p. 319.

5. Simplicius *Physics* 176.29, 175.12–14. See Richard D. McKirahan, *Philosophy before Socrates* (Indianapolis/Cambridge: Hackett Publishing, 2010), p. 194 (Kindle ed.).

6. Sextus Empiricus *Against the Professors* 7.140, trans. Demetris Nicolaides. See also Graham, *Texts of Early Greek Philosophy*, p. 309 (text 63).

7. Diogenes Laertius 2.6–15, trans. Demetris Nicolaides. See also Graham, *Texts of Early Greek Philosophy*, p. 275 (text 1).

8. Ibid.

9. Ibid.

CHAPTER 17. DEMOCRITUS AND ATOMS

1. Sextus Empiricus *Against the Professors* 7.135, trans. Erwin Schrödinger, *Nature and the Greeks and Science and Humanism* (Cambridge: Cambridge University Press, 1996), p. 89.

2. Plutarch *Against Colotes* 1110f–1111a, trans. Daniel W. Graham, *The Texts of Early Greek Philosophy: The Complete Fragments and Selected Testimonies of the Major Presocratics* (Cambridge: Cambridge University Press, 2010), p. 537 (text 28).

3. Aristotle *On Generation and Corruption* 315b6–15. See Graham, *Texts of Early Greek Philosophy*, p. 541 (text 41).

4. Ibid., 324b35–325a6, a23–b5, trans. Graham, *Texts of Early Greek Philosophy*, p. 529 (text 14).

5. Sextus Empiricus *Outlines of Pyrrhonism* 2.63, trans. Demetris Nicolaides. See Greek book Βας. Α. Κύρκος, *Οι Προσωκρατικοί: Οι Μαρτυρίες και τα Αποσπάσματα τόμος Β* (Αθήνα: Εκδόσεις Δημ. Ν. Παπαδήμα, 2007), p. 255.

6. Graham, *Texts of Early Greek Philosophy*, pp. 579–95.

7. Sextus Empiricus *Against the Professors* 7.138–139, trans. Graham, *Texts of Early Greek Philosophy*, p. 597 (text 140).

8. Richard P. Feynman, *The Feynman Lectures on Physics* (Boston: Addison-Wesley Publishing Co., Inc., 1963), p. 1-2.

9. Leon Lederman and Dick Teresi, *The God Particle: If the Universe Is the Answer, What Is the Question?* (Boston: Houghton Mifflin, 1993), p. 340.

10. Aristotle *Metaphysics* 985b4–20. See Graham, *Texts of Early Greek Philosophy*, p. 525 (text 10).

11. Simplicius *On the Heavens* 242.15–26, trans. Graham, *Texts of Early Greek Philosophy*, p. 533 (text 23).

12. Bertrand Russell, *The History of Western Philosophy* (New York: Simon & Schuster, 1945), p. 71.

13. Plutarch *Against Colotes* 1108f–1109a, p. 527 (text 13).

14. Isaac Asimov, *Understanding Physics* (US: Dorset Press, 1988), p. 7.

15. Lederman and Teresi, *The God Particle*, p. 44.

16. Banesh Hoffman, *The Strange Story of the Quantum* (New York: Dover Publications Inc., 1959), p. 68.

17. Brian Greene, *The Elegant Universe: Superstrings, Hidden Dimensions, and the Quest for the Ultimate Theory* (New York: W. W. Norton & Company, 1999), p. 114.

18. Jeffrey Bennett, Megan Donahue, Nicholas Schneider, and Mark Void, *The Essential Cosmic Perspective 7th Edition* (Boston: Pearson, 2013), p. 479.

19. Werner Heisenberg, *Physics and Philosophy: The Revolution in Modern Science* (New York: Harper Torchbooks, 1962), p. 40.

20. Aëtius 1.3.18, S 1.14.1f. See Graham, *Texts of Early Greek Philosophy*, p. 537 (texts 31, 32); Cicero *On Fate* 20.46. See Graham, *Texts of Early Greek Philosophy*, p. 537 (text 33).

21. Aëtius 1.4.1–4. See Graham, *Texts of Early Greek Philosophy*, pp. 541–45.

22. Sextus Empiricus *Against the Professors* 7.135, p. 89.

23. Galen *On Medical Experience* 15.7, trans. Demetris Nicolaides. See also Graham, *Texts of Early Greek Philosophy*, p. 597 (text 139).

24. Sextus Empiricus *Against the Professors* 7.138–139, p. 597 (text 140).

25. Ibid., 7.140, trans. Demetris Nicolaides. See also Graham, *Texts of Early Greek Philosophy*, p. 309 (text 63).

26. "Astronomers Conclude Habitable Planets Are Common," Institute for Astronomy University of Hawaii, http://www.ifa.hawaii.edu/info/press-releases/HabitablePlanets Common/ (accessed March 23, 2014).

27. Bennett, *Essential Cosmic Perspective*, p. 511; Carl Sagan, *Cosmos* (New York: Random House, 1980), p. 300.

BIBLIOGRAPHY

Asimov, Isaac. *Asimov's Chronology of Science and Discovery*. New York: HarperCollins, 1989.

———. *Asimov's Chronology of the World*. New York: HarperCollins, 1991.

———. *The Greeks; A Great Adventure*. Boston: Houghton Mifflin, 1965.

———. *Understanding Physics*. US: Dorset Press, 1988.

Baker, Joanne. *50 Physics Ideas You Really Need to Know*. UK: Quercus, 2007.

Bertman, Stephen. *The Eight Pillars of Greek Wisdom*. New York: Barnes & Noble, 2007.

———. *The Genesis of Science: The Story of Greek Imagination*. Kindle ed. Amherst, NY: Prometheus Books, 2010.

Boardman, John, Jasper Griffin, and Oswyn Murray, eds. *The Oxford Illustrated History of Greece and the Hellenistic World*. Oxford: Oxford University Press, 1986.

Brunschwig, Jacques, and Geoffrey E. R. Lloyd. *A Guide to Greek Thought: Major Figures and Trends*. Cambridge, MA: Belknap Press of Harvard University Press, 2003.

———. *Greek Thought: A Guide to Classical Knowledge*. Cambridge, MA: Belknap Press of Harvard University Press, 2000.

Burckhardt, Jacob. *The Greeks and Greek Civilization*. Edited by Oswyn Murray. Translated by Sheila Stern. New York: St. Martin's Griffin, 1999.

Burkert, Walter. *Greek Religion: Archaic and Classical*. Translated by John Raffan. MA: Blackwell Publishing Ltd. and Harvard University Press, 1985.

Burnet, John. *Early Greek Philosophy*. London: A & C Black, 1920.

Dalling, Robert. *The Story of Us Humans, From Atoms to Today's Civilization*. New York, Lincoln Shanghai: iUniverse, 2006.

Davies, P. C. W., and Julian Brown. *Superstrings: A Theory of Everything?* Cambridge: Cambridge University Press, 1992.

Diamond, Jared. *The Third Chimpanzee*. New York: HarperCollins, 1992.

Economou, Eleftherios N. *A Short Journey from Quarks to the Universe*. Berlin and Heidelberg: Springer, 2011.

Einstein, Albert. *Relativity: The Special and the General Theory*. Kindle ed. Amazon Kindle Direct Publishing, 2011.

Feynman, Richard P. *The Meaning of It All*. New York: Basic Books, 1998.

———. *Six Easy Pieces*. New York: Perseus Publishing, 1963.

———. *Six Not So Easy Pieces*. New York: Perseus Publishing, 1963.

Freeman, Charles. *The Greek Achievement: The Foundation of the Western World*. New York: Penguin Books, 2000.

Freeman, Kathleen. *Ancilla to the Pre-Socratic Philosophers*. Cambridge, MA: Harvard University Press, 1996.

Graham, W. Daniel. *Explaining the Cosmos: The Ionian Tradition of Scientific Philosophy*. NJ: Princeton University Press, 2006.

Graham, W. Daniel, ed. *The Texts of Early Greek Philosophy: The Complete Fragments and Selected Testimonies of the Major Presocratics*. Translated by W. Daniel Graham. Cambridge: Cambridge University Press, 2010.

Graves, Robert. *The Greek Myths*. Canada: Penguin Group, 1955.

Greene, Brian. *The Elegant Universe: Superstrings, Hidden Dimensions, and the Quest for the Ultimate Theory*. New York: W. W. Norton & Company, 1999.

———. *The Fabric of the Cosmos: Space, Time, and the Texture of Reality*. New York: Vintage, 2005.

———. *The Hidden Reality: Parallel Universes and the Deep Laws of the Cosmos*. New York: Vintage, 2011.

Gregory, Andrew. *Eureka! The Birth of Science*. Cambridge: Icon Books, 2001.

Hamilton, Edith. *The Greek Way*. New York and London: W. W. Norton & Company, 1930.

Hardy, Alister. *The Biology of God*. New York: Taplinger Publishing, 1976.

Hawking, Stephen. *A Brief History of Time: From the Big Bang to Black Holes*. New York: Bantam Books, 1988.

Heisenberg, Werner. *Physics and Philosophy: The Revolution in Modern Science*. New York: Harper Torchbooks, 1962.

James, Renée C. *Seven Wonders of the Universe: That You Probably Took for Granted*. Baltimore: Johns Hopkins University Press, 2011.

Kirk, G. S., J. E. Raven, and M. Schofield. *The Presocratic Philosophers*. Cambridge: Cambridge University Press, 1983.

Kolac, Daniel, and Garrett Thomson. *The Longman Standard History of Philosophy*. New York: Pearson, 2005.

Lederman, Leon, and Dick Teresi. *The God Particle: If the Universe Is the Answer, What Is the Question?* Boston: Houghton Mifflin, 1993.

Lightman, Alan. *Great Ideas in Physics*. 3rd. ed. New York: McGraw-Hill, 2000.

Lindberg, David C. *The Beginnings of Western Science: The European Scientific Tradition in Philosophical, Religious, and Institutional Context, Prehistory to A.D. 1450*. 2nd. ed. Chicago: University of Chicago Press, 2008.

Lloyd, G. E. R. *Early Greek Science: Thales to Aristotle*. New York: W. W. Norton & Company, 1970.

———. *Greek Science after Aristotle*. New York: W. W. Norton & Company, 1973.

McKirahan, Richard D. *Philosophy before Socrates*. Indianapolis/Cambridge: Hackett Publishing, 2010.

Menzies, Allan. *History of Religion: A Sketch of Primitive Religious Beliefs and Practices, and of the Origin and Character of the Great Systems*. Memphis, TN: General Books, 2010.

Mourelatos, Alexander P. D., ed. *The Pre-Socratics: A Collection of Critical Essays*. Garden City, NY: Doubleday and Company, 1974.

Oppenheimer, Robert J. *Science and the Common Understanding*. New York: Simon & Schuster, 1954.

Pomeroy, Sarah B., Stanley M. Burstein, Walter Donlan, and Jennifer Tolbert Roberts. *A Brief History of Ancient Greece: Politics, Society, and Culture*. 2nd. ed. Oxford: Oxford University Press, 2008.

Popper, Karl R. *Conjectures and Refutations: The Growth of Scientific Knowledge*. London and New York: Routledge, 1989.

Randall, Lisa. *Knocking on Heaven's Door: How Physics and Scientific Thinking Illuminate the Universe and the Modern World*. New York: Harper Perennial, 2011.

————. *Warped Passages: Unraveling the Mysteries of the Universe's Hidden Dimensions*. New York: Ecco, 2005.

Ridley, B. K. *Time, Space and Things*. 2nd. ed. Cambridge: Cambridge University Press, 1984.

Rosenblum, Bruce, and Fred Kuttner. *Quantum Enigma: Physics Encounters Consciousness*. 2nd. ed. Oxford: Oxford University Press, 2011.

Russell, Bertrand. *The History of Western Philosophy*. New York: Simon & Schuster, 1945.

————. *Religion and Science*. New York and Oxford: Oxford University Press, 1997.

————. *The Scientific Outlook*. London and New York: Routledge, 2009.

Sagan, Carl. *Cosmos*. New York: Random House, 1980.

Schrödinger, Erwin. *Nature and the Greeks and Science and Humanism*. Cambridge: Cambridge University Press, 1996.

————. *What Is Life? & Mind and Matter*. Cambridge: Cambridge University Press, 1967.

Sean, Carroll. *The Particle at the End of the Universe*. New York: Dutton, 2012.

Sherrington, Charles. *Man on His Nature*. Reissue. Cambridge: Cambridge University Press, 2009.

Stark, Rodney. *Discovering God: The Origins of the Great Religions and the Evolution of Belief*. Kindle ed. HarperCollins, 2009.

Tattersall, Ian. *The World from Beginnings to 4000 BCE*. Oxford: Oxford University Press, 2008.

Thornton, Bruce. *Greek Ways: How the Greeks Created Western Civilization*. San Francisco: Encounter Books, 2002.

Vlastos, Gregory. *Studies in Greek Philosophy*. Vol. 1: The Presocratics. Princeton: Princeton University Press , 1993.

Waterfield, Robin. *The First Philosophers: The Presocratics and Sophists*. Oxford and New York: Oxford University Press, 2000.

Wrangham, Richard. *Catching Fire: How Cooking Made Us Human*. New York: Basic Books, 2009.

INDEX